God's Two Books

Copernican Cosmology and Biblical Interpretation in Early Modern Science

KENNETH J. HOWELL

UNIVERSITY OF NOTRE DAME PRESS
Notre Dame, Indiana

Manufactured in the United States of America

Library of Congress Cataloging-in-Publication Data
Howell, Kenneth J. (Kenneth James)
    God's two books : Copernican cosmology and biblical interpretation
in early modern science / Kenneth J. Howell.
        p.   cm.
Includes bibliographical references and index.
    ISBN 0-268-01045-5 (cloth : alk. paper)
    1. Religion and science.   I. Title.
    BL240. .H76 2001
    261.5'5'09409031—dc21   2001004287

∞ *This book was printed on acid-free paper.*

Contents

Preface

Questions concerning the interface of the empirical sciences and theology are receiving renewed attention in our day. In the last twenty-five years numerous institutes and conferences have addressed many aspects of this interaction. Although these discussions have arisen because of changing theories in science, many of the more fundamental problems date back to antiquity with a long and venerable history. Whatever the future of science and theology may be, a knowledge of their respective histories is essential for reflection and theorization.

Progress in writing history results from frequent reinterpretation of the past, and the history of science as a discipline has matured greatly in the last fifty years through historicization, a process that attempts to understand great figures of past science through categories they employed and espoused rather than those imposed by the present. With this in mind, I have attempted to explicate the complex interactions between astronomy, cosmology, and biblical interpretation in the early modern period through contextualizing the thinking of the historical participants from within their own historical frameworks.

This history may interest scholars from a wide range of disciplines: historians of science, church historians, hermeneutical theorists, as well as all students of early modern intellectual history. The Renaissance saw an intellectual vibrancy the likes of which have rarely been equaled. The primary sources from this period call forth from the historian deep and diverse interpretative skills. It is my hope that many scholars and students may gain a deeper understanding of how rich was the interaction of biblical interpretation and cosmology in a period of history very different from our own.

An author incurs many debts in writing a book of this kind. The Pascal Centre for Advanced Studies in Faith and Science, established in 1988 by Redeemer College, is the institution most responsible for financial and research support. The Centre's open-minded investigation into issues of faith and science speaks well of its contributions to science studies. Its founding director, Professor Jitse van der Meer, has been a constant encouragement to me. Still, the conclusions in this book are entirely my own responsibility. Professor John Hedley Brooke, formerly of Lancaster University and now of Oxford University, has been the leading intellectual impetus behind this research. Few in the history of science profession can match his gentlemanly and scholarly stature. Professors Nicholas Jardine (Cambridge) and Stephen Pumfrey (Lancaster) offered invaluable suggestions and encouragements. My thanks goes to Professor Jardine for his permission to use Chapter 3, the substance of which appeared in *Studies in the History and Philosophy of Science*. The Institute for the History of Science in Utrecht provided me with the means to investigate the Dutch sources used in Chapter 5. Among my American friends, the Institute for Advanced Study, the Lilly Library, and the Department of the History and Philosophy of Science at Indiana University provided a scholarly home during much of its writing. The guidance and personal friendship of the late Richard S. Westfall offered me a paternal concern that transcended the bounds of scholarly acquaintance; Sam will be sorely missed. Monsignor Dr. Stuart Swetland, chaplain and director of the Newman Foundation at the University of Illinois, provided financial support during the final stages of editing. Professor Robert Prescott of Bradley University offered numerous modifications and encouraged me to believe that this book will be of interest to scholars outside the history of science. I wish to thank Lance Magnotta and Rebekah Howell, my research assistants, and Alina Willis, my Administrative Assistant, for their diligent help in the final editing. Thanks also belongs to James Langford, director of the University of Notre Dame Press, and to anonymous referees, for their suggestions.

Early Modern Science and Biblical Interpretation

*A*ristotle pointed out that everyone desires to know by nature, but clearly not all humans agree on how to know. Do inborn and rational thoughts provide the path to knowledge, or is experience in the sensible world the primary means to acquire knowledge? Or does knowledge come from a supernatural source by special acts of divine disclosure? When these different sources of knowledge conflict with one another, what strategies should be used to settle these conflicts? Are certain established disciplines in a privileged position to adjudicate between knowledge claims or are all on equal ground? Such questions do indeed seem to press upon us today with as much or more force than they did in ancient Greece. Although we may question the universality of Aristotle's dictum, we are nevertheless struck by the repeated and perennial character of the questions.

This book is the story of some major figures of early modern Europe who wrestled with these very questions as they faced dramatic changes in their conceptions of the universe between the dawn of the sixteenth century and the close of the seventeenth. It is the story of conflicts between various sources of knowledge and belief, all of which were held to be essential to a proper understanding of the nature of things. It is the story of astronomers in conflict and cooperation, deep in thought and keen in observation, who believed that God could be known in both nature and the Bible and that God had entrusted the task of interpretation of both realms to the human family.

In 1543 Nicholas Copernicus published *On the Revolutions of the Heavenly Spheres*, the fruit of his lifelong labors in astronomy. The fallout from that book was to have profound effects on Europeans' conception of the universe. Copernicus revived the ancient system put forth by Aristarchus that placed the sun at the center of the known universe. Copernicus's attribution of triple motion to the earth raised deep problems for physics and theology because his claims went counter to the prevailing Aristotelian physics of motion and to the common interpretation of certain biblical texts (e.g., Psalm 93, Joshua 10) which seemed to teach that the earth could not move. The situation was further complicated by different interpretations of the status of astronomy as a discipline. Some in this period (e.g., Osiander, Bellarmine) relied on the ancient view that astronomy had fulfilled its role when it properly predicted the motions of celestial objects. There was no presumption that astronomy could lead to truth about the universe. The astronomer had done his job when he "saved the phenomena." Others, such as Kepler and Galileo, believed that astronomy (and the mathematics it employed) was a key to the true constitution of the universe:

> Philosophy is written in that grand book, the universe, which stands continually open to our gaze. But it cannot be understood unless one first learns to comprehend the language and read the letters in which it is written. It is written in the language of mathematics, and the characters are triangles, circles, and other geometrical figures, without which it is not humanly possible to comprehend a single word.[1]

Those who adopted this realist view of astronomy incurred the additional task of giving physical explanations of astronomical theory. By the end of the seventeenth century, an alternative physics, developed primarily by Galileo and Newton, was well in place, and this new physics provided additional explanations of the Copernican astronomy. Almost all the participants in these developments believed in the authority of the Bible, and many of them embraced historic orthodox Christianity. Not all, however, agreed on how to interpret the Bible, and they were divided over whether the notion of a moving earth could be reconciled with biblical texts that seemed to suggest otherwise. The goal of these realist astronomers and natural philosophers was increasingly to arrive at and to expound the true system of the universe. As this shift from an instrumental to a realist view of astronomy occurred, many astronomers faced the empirical problem of the celestial spheres, a problem attacked from the standpoints of astronomy, physical theory, and biblical interpretation. The result was a gradual relinquishing of the notion of solid spheres that provoked a reevaluation of theories of celestial matter and of the relevance of the Bible.

COPERNICANISM AND BIBLICAL INTERPRETATION

The role of the Bible with respect to changes in cosmology during the early modern period is a subject with a long pedigree. Copernicanism and biblical interpretation are standard subjects in their respective professional disciplines, the history of science and the history of theology. The interactions between them, however, have largely fallen to historians of science who have expended considerable energy on the effects of the new cosmology on biblical interpretation. Most of the literature on Copernicanism and the Bible has dealt with Catholic Europe and the Galileo affair. Sometimes the impression is conveyed that not much remains to be said.

I will argue that we should not be too confident in our knowledge of the role of the Bible in the Copernican debates because that knowledge is both incomplete and inaccurate. Our understanding of the fate of Copernicanism in the Protestant regions of northern Europe has grown considerably in the last thirty-five years, but with little corresponding scholarship on the theological aspects of those developments. This lacuna contrasts sharply with the abundance of primary sources from the period that have remained unanalyzed and unsynthesized in a larger historical framework. This book fills that void by giving careful attention to the writings of astronomers and theologians in Protestant northern Europe that dealt with the cosmological implications of biblical texts. As we shall see, this research provides a larger context in which to understand the Catholic debates, and it has profound implications for understanding the interactions of science and religion during the sixteenth and seventeenth centuries. Counter to the trends in historical scholarship, I will focus not on the effects of the new cosmology on biblical hermeneutics but on the varied roles played by biblical interpretation in the emerging cosmologies of early modern science.

Since religion in early modern Europe necessarily entailed the Christian religion and since Christianity was a religion of the Book, the problem of conflict between Scripture and the Copernican theory was bound to arise. The note of inevitability rang loudly in the warfare model of science and religion advanced by the nineteenth-century American historians John Henry Draper and Andrew Dickson White.[2] For them, the only function of the Bible in the Copernican debates was a negative one, the repression of new ideas by clinging to interpretations that took precedence over science.[3] The warfare historians had a plethora of evidence to buttress their claims. Luther's famous *Table Talks* showed him to be opposed to the novelty of the new astronomy. The citation of Joshua 10, implying the motion of the sun and the stability of the earth, was sufficient in his mind to settle the issue: Scripture right, Copernicus wrong. Although his protégé Philipp Melanchthon was milder and less combative than Luther, he too was no less committed to the traditional conjunction of the Ptolemaic-Aristotelian system and biblical authority. Consequently, Melanchthon argued for the full acceptance of scriptural declarations over some unproven theory in astronomy.[4] Nor did the Reformed (Calvinist) side of the Reformation fare much better. With a strong commitment to the literal authority of biblical texts, John Calvin of Geneva denounced the new astronomer as opposed to Scripture ("Who will venture to place the authority of Copernicus above that of the Holy Spirit?").[5] According to the warfare model, what the magisterial Reformers started, the lesser lights of later generations continued. The names of Turretin, John Owen, Hutchinson, Pike, Horsley, and many others carried the opposition to heliocentrism down into the nineteenth century. As for astronomers and physicists who were deeply religious and trained in their traditions, a few such as Rheticus and Reinhold were convinced of the Copernican theory but were readily silenced and suppressed.

According to White's warfare metaphor, the most visible opposition to Copernicanism, of course, came in the form of a war waged by the Roman Catholic Church against Galileo. Galileo's dismissal of the rule of Scripture over science made him worse than Luther or Calvin in the eyes of churchmen and populace alike. The fate of Galileo stood as a reminder of the ill effects that follow from ecclesiastical control over the pursuit of scientific truth.[6] For White especially, the virulent opposition to Copernicanism was unmistakable evidence for the intrinsic incompatibility of dogmatic theology and scientific knowledge. Religion, which consists of feeling and which makes no claims of cognitive truth, is not seen as intrinsically antiscientific. But dogmatic and polemic theology, with its reliance on authoritative texts, produced a mind-set which led to the eradication of the true knowledge of nature.[7]

In the post–World War II growth of the history of science as a discipline, many scholars arrived at a deeper understanding of the scientific revolution. The work of A. Rupert Hall and Marie Boas Hall, Anneliese Meyer, Alexandre Koyré, Richard Westfall, and others painted a more complex and nuanced picture of the development of the sciences. One result was the awareness that opposition to Copernicanism was not solely theological but that there were also bona fide scientific obstacles to its acceptance. Nevertheless, older views on the relation of Copernicanism to the Bible remained largely unaffected. Although Koyré in 1961 was ready to correct White's overstatements about biblical authority being the predominant obstacle to scientific progress, he nevertheless admitted that "Biblical authority was a powerful obstacle to the diffusion of the heliocentric theory, and prevented its general acceptance until the beginning of the eighteenth century."[8] These studies, then, placed the use of biblical authority in a larger context of opposition to Copernicanism, but the analyses still remained at a very general level. The Bible still found no positive function in the cosmological disputes emerging in the late sixteenth century, and the only hermeneutical strategy that received significant attention was the Copernicans' appeal to accommodation.[9]

Around the same time, increased research into the Galileo affair produced an avalanche of studies that bore on Galileo's battle with the Roman Catholic Church.[10] Interpretations of Galileo's encounter with the church are as bewildering as they are diverse, but the hermeneutical nuances of biblical interpretation involved in this affair have received

surprisingly little attention until recently. Pedersen suggested that the
Galileo affair should be studied as an episode in the history of theology
as well as of science. He argued that the church's condemnation of
Copernican astronomy was best explained by the role played by the
Council of Trent in stressing the priority of literal interpretation of
the Bible. He stressed also the unusually rapid way in which the decision was made, a process that normally would have taken much
longer.[11] Placing the affair into the history of theology has fallen to
Blackwell, whose study represents the most thorough treatment of the
role of the Bible in Galileo's two encounters with church authorities.
Blackwell argues that the Galileo affair resulted from the growing cen-
tralization of authority within the church. The trial of 1633 was only
tangentially about science and religion and more about obedience to
the authority of the church.[12] This authority was linked with the concern of the church to remain the guardian of interpretation. Blackwell
sees the prohibition against private interpretation defined at the fourth
session of the Council of Trent as decisive for Galileo's fate.

The role of the Bible in the Copernican debates has been dominated by the Galileo affair, but the issues were prominent in the
thought and writings of other thinkers in this period. Like his Italian
counterpart, Kepler argues for accommodation in his *Introduction* to
the *Astronomia Nova* when he systematically attempts to answer objections to Copernicanism.[13] Similar arguments can also be found in the
writings of the seventeenth-century Anglican cleric John Wilkins, who
invoked accommodation as a necessary feature of the Bible.[14] Scholarly
accounts of these strategies have often remained at a superficial level.
Even those who disavow a warfare or strict-separation account of science and religion in the early modern period still pose the Copernican
debates in terms of literal (traditional) versus nonliteral (innovative)
interpretations of the Bible.[15] Blackwell's analysis refines the notion of
literal interpretation into strictly literal and metaphorical meanings, a
procedure that gives room to less physical meanings of biblical words.
Blackwell sees Cardinal Bellarmine's theory of *de dicto* biblical inspiration as the conceptual background that allowed him to contend *contra*
Galileo and Foscarini that everything mentioned in the Bible was infallible since it was spoken by the One who cannot lie. Thus physical

questions, too, were judged matters of faith and morals unless demonstrated to be otherwise.[16] Blackwell has pointed us in the right direction, but I will also argue that the role of the Bible in the cosmologies of the Copernicans is more varied and nuanced than simply the invocation of accommodation. A more careful comparative analysis of hermeneutical arguments used in the Copernican debates reveals that Galileo's approaches to hermeneutics are even less innovative than we imagined, since a similar variety of approaches can be found in other primary sources that could not possibly have been influenced by Galileo directly.[17]

The only recent book-length study of Protestant biblical interpretation in the emergence of early modern science comes from Peter Harrison. Harrison's central thesis contends that the Protestant Reformers' insistence on literal interpretation played a central though not exclusive role in opening up the possibility of a "new conception of the order of nature."[18] Because Protestant interpretation eliminated "the capacity of things to act as signs" (i.e., allegory), he argues, "the study of the natural world was liberated from the specifically religious concern of biblical interpretation, and the sphere of nature was opened up to new ordering principles."[19] Reversing the common view that the new sciences provoked new interpretations of the Bible, Harrison maintains that sixteenth-century people "found themselves forced to jettison traditional conceptions of the world" when they "began to read the Bible in a different way."[20] His excellent sketch of the history of hermeneutics shows that prior to the sixteenth century the fourfold system of interpretation allowed a wide variety of readings that were abandoned under the impetus of the Protestant insistence on the author's intention. This abandonment of the *quadriga* meant that "the text of scripture was for the first time exposed to the assaults of history and science."[21] Harrison generalizes his historical assessment by claiming that "the transformations which brought on the birth of modernity moved western culture from the era of 'the two books' to that of 'the two cultures.'"[22] While Harrison seems aware of many seventeenth-century figures who still viewed nature through the lenses of "the old symbolic order," he argues that these thinkers are "indicative of an unconscious reluctance to admit the failure of the old world picture."[23]

The merits of Harrison's study are many, but his fundamental thesis fails to account for the diversity of hermeneutical approaches that we will observe in the following chapters. Although he includes a wide range of primary sources, he rarely treats many of the sources used here, or gives them only a cursory reading.[24] His commentary on Kepler is limited to the appeal to accommodation found in the *Astronomia Nova*.[25] As I show in chapter 4, Kepler's interpretation is much wider and deeper. And Kepler, a committed Lutheran, can hardly be said to have given up a symbolic view of nature. His Trinity-sphere icon rivals the most allegorical view of the cosmos. For Kepler, the whole universe contained signs of the Creator, and Philipp Lansbergen, a Calvinist minister, espoused a view similar to Kepler's. Both were embued with a Protestant hermeneutic and were strong advocates of the new astronomy, but they also saw meaning in the universe on several different levels. Harrison also seems unaware of the Domincan Tommaso Campanella's arguments that Galileo's cosmology fulfilled a literal meaning of Scripture. Harrison has contributed to a clearer understanding of many primary sources, especially English ones, but his overall thesis ignores much exegetical diversity in the interaction of science and interpretation in early modern Europe.

THE RECEPTION OF COPERNICANISM

One important result of the reevaluation of the scientific revolution during the last few decades has been an enriched understanding of the reception of Copernicanism in various national contexts.[26] The standard view of the Copernican revolution derived from the work of mid-century historians stressed its discontinuity with medieval science by underscoring the empirical anomalies and complexities of medieval astronomy. These cumbersome complexities were replaced by the simplicity of Copernicus's achievement, which resulted from the growing crisis of astronomy in the fifteenth century. The motivating force behind that achievement was a growing mathematical realism adopted by all subsequent Copernicans. This mathematization of nature argued that God made the world according to measurement, weight, and number so that only through mathematics could the secrets of the world system be unveiled. Since relatively few Copernicans can be found in the sixteenth century between *De Revolutionibus* (1543) and Kepler's *Mysterium Cosmographicum* (1596), it is no surprise that there was not any development of Copernican cosmology. This led Bernard Cohen to argue that there was no Copernican revolution in astronomy until ca. 1600 because Copernicus's work did not significantly alter the manner in which astronomers did their work (e.g., planetary motions, relative distances, etc.), or bring about any significant change in cosmology (e.g., agreement on celestial matter). For Cohen, the real revolution came with Kepler because he sought physical causes of planetary movements.[27]

In the last thirty years this standard picture has undergone significant modification.[28] It seems certain that the full story of the spread of Copernicanism throughout Europe is still to be told, but several features have already altered our view of what was thought to be a well understood segment of history. Among these, one can count at least three features that have led to major historical revisions: the role of aesthetics in the emergence of the new astronomy, the Wittenberg interpretation of Copernicus, and a more careful knowledge of attempts to combine geocentric and heliocentric systems in the late sixteenth century.

In a groundbreaking article, Owen Gingerich challenged the view that Copernicus himself faced mounting empirical anomalies. Gingerich argued that Copernicus sought the "fixed symmetry of its [universe's] parts" with his grand aesthetic vision. That cosmological vision transformed "the monster of Ptolemy" into "the man of Copernicus," a system in which the relative sizes of planetary orbits were fixed with respect to each other and could no longer be scaled in size.[29] Although the focus in this book is on the reception rather than the creation of Copernicanism, the continuing relevance of the aesthetic search for symmetries will prove to be significant in understanding the theological aspects of the Copernican debates. Kepler's geometrical vision of the cosmos is only the most prominent example. We shall observe that aesthetic considerations were also central to Tycho's search for the restoration of astronomy.

Research on Copernicanism in northern Protestant Europe has also shown a widespread knowledge of *De Revolutionibus* among the teachers and students of astronomy in the sixteenth century. The work of Dobrzycki, Gingerich, Westman, and others has shown a lineage of annotations to copies of *De Revolutionibus* that:

> reveal that even if Copernicus's revolutionary new doctrine failed to find a place in the regular university curriculum, a network of astronomy professors scrutinized the text and their protégés carefully copied out their remarks, setting the notes onto the margins of fresh copies of the book with a precision impossible by aural transmission alone. Clearly the students sat with the master book before them as they transcribed key words, data, diagrams or whole paragraphs of elucidation.[30]

A key institution in this process of transmission was the center of the Lutheran Reformation, the University of Wittenberg. The Wittenberg interpretation of Copernicus developed under the influence of Luther's right-hand man, Philipp Melanchthon, who became the *Praeceptor Germaniae* because of the extensive educational system that emerged under his influence. Melanchthon's encouragement of astronomy at Wittenberg and other universities played a singular role in disseminating the knowledge of the Copernican theory.[31] Key figures in this process were Erasmus Reinhold, Georg Joachim Rheticus, Caspar Peucer, Michael Mästlin, and Johannes Kepler. Reinhold's use of Copernican parameters in his *Prutenic Tables* (1551) indicates how the circle around Melanchthon was taking the theory. While willing to use all possible observational and mathematical data for more accurate prediction, this group of astronomers made no attempt to judge the cosmological truth of *De Revolutionibus*.[32] This "split interpretation" (separating cosmology from predictive astronomy) became a tool for teaching and refining the Copernican theory. The Wittenberg circle represented a line of interpretation that played a significant role in the theoretical development of heliocentrism as well as evoking a variety of theological evaluations.

The influence of Wittenberg led also to the development of geo-heliocentric models of the universe in the late sixteenth century. When some astronomers trained in the spirit of Wittenberg did finally address the cosmological implications of Copernicus, the physical and theological arguments seemed to require a major modification of both Ptolemy and Copernicus. Perhaps the central but by no means only figure in this story was Tycho Brahe, the greatest observational astronomer in the West prior to the invention of the telescope. As the sixteenth century wore on, the problem of celestial matter and the existence of celestial spheres came to the fore as appearances like the nova of 1572 and the comets of 1577 and 1585 caused astronomers to rethink the traditional solid spheres. Between ca. 1575 and 1650 radical reassessments of celestial matter were introduced that were as revolutionary as Copernicus's original proposal.[33]

With a deeper appreciation of the cosmological context of Protestant biblical interpretation, it becomes possible to reassess the differences between Catholic and Protestant hermeneutics of the early modern period. Similar strategies show up on both sides of the ecclesiastical divide, but these functioned in distinct ecclesiastical contexts. Both Protestants and Catholics inherited a common heritage of ancient Christian interpretation, but they exercised their exegetical skills in very different environments, contexts that had profound implications for how they couched their arguments.

The central argument of this book contends that the use of the Bible in early modern cosmology is considerably more complex and subtle than has heretofore been recognized, a use that cannot be easily categorized into a Copernican/non-Copernican dichotomy. The common notion that the Bible functioned mainly as a deterrent to the acceptance of Copernicanism is very wide of the mark because both Copernicans and non-Copernicans viewed the Bible as offering truth about the physical universe, albeit in different ways. The appeal to accommodation on the part of the Copernicans was recognized also by non-Copernicans, but from this recognition they drew different conclusions. For neither, however, did the recognition of accommodated language in the Bible mean its complete irrelevance for cosmology. An adequate analysis of the hermeneutics involved requires a more refined understanding of literal interpretation, a notion that is not unitary or uncomplicated. What appears most important is the notion of truth, especially mathematical truth. The problems of terrestrial motion and celestial matter would not have taken on the proportions that they did if the common assumption of mathematical truth and biblical truth had not forced the early moderns into an unprecedented struggle to reconcile information from science with biblical data.

Writing history requires a constant shedding of our misconceptions about the past. Our understanding may be revised by new documents that come to light or by new interpretations that cast old material in a new light. The theological aspects of the Copernican debates, once thought to be well understood, yield to both types of reinterpretation. The panoply of documents studied in this book requires a new intellectual taxonomy that draws its lines in finer detail than before. The Bible was used not only to limit and question various cosmologies but to ground and support them as well. That same Bible was read in a variety of ways that cannot be classified into a simple literal vs. non-literal dichotomy because the hermeneutical awareness of the historical participants was too subtle and complex to yield to such a categorization. What that new taxonomy ultimately looks like will only emerge after more of the primary documents are studied and more subtle analyses are offered. What my analysis shows is that the strategies of the early moderns for pursuing cosmological and theological truth derived from antiquity and remained robust up to their day.

CHAPTER ONE

Reading the Heavens and Scripture in Early Modern Science

*W*hen Copernicus dedicated *De Revolutionibus orbium coeles-tium* to the "Most Holy Father" (Pope Paul III) in 1543, he spoke of his labors as an explication of the world machine that "the Best and Most Orderly Workman" had constructed "for our sake."[1] Within his short preface, the canon of Frauenberg united three areas of life central to readers of the sixteenth century: religion, humanity, and nature. If the first two of this triad were to be sustained, the last member had also to be advanced. These concerns remained unabated at the end of the seventeenth century when John Ray, the English cleric and natural philosopher, penned *The Wisdom of God Manifested in the Works of Creation* (1691). Ray, no less than hundreds of writers before him, marveled in the sheer diversity of creation as a manifestation of the magnanimous mercy of the Creator, and he contemplated the finely tuned structure of the world as a display of the wisdom of God.[2] The scientific revolution, once portrayed by historians as a driving force in the secularization of Western culture, occurred in a milieu so imbued with religious devotion and conviction that any investigation into the heavens, or into nature more generally, required constant reflection on the relation of natural knowledge to divine revelation and doctrinal belief. As has been noted many times, one of the primary modes of

Girolamo Fracastoro,
from his *Homocentria*,
1538 edition. Courtesy
of The Lilly Library,
Indiana University,
Bloomington, Indiana.

HIERONYMI FRACASTORII

justifying natural inquiry in the early modern period was its religious utility, a stratagem evident in Thomas Sprat's apologetic history of the Royal Society:

> They both [science and Protestantism] may lay equal claim to the word Reformation; the one having compassed it in Religion, the other purposing it in philosophy. They both have taken a like course to bring this about; each of them passing by the corrupt copies, and referring themselves to the perfect Originals for their instruction; the one to the Scripture, the other to the huge Volume of Creatures. They are both accused unjustly by their enemies of the same crimes, of having forsaken the Ancient Traditions, and ventured on Novelties. They both suppose alike that their Ancestors might err; and yet retain a sufficient reverence for them. They both follow the great Precept of the Apostle, of trying all things. Such is the harmony between their interests and tempers.[3]

At the outset of Sprat's century, Francis Bacon argued strongly for a certain degree of separation of religion and science from his conviction that true theories of nature could emerge only from intensely empirical inquiry. For Bacon, the forces of the English church were often misguided in their efforts to constrict an empirical study of nature by a spurious appeal to authority. At the same time, Bacon believed that proper science would reinforce religion. In his allegorical treatment of ancient myths, Bacon interpreted Pan (meaning the universe) as an "allegory plainly divine, seeing that next to the Word of God, the image itself of the world is the great proclaimer of the divine wisdom and goodness."[4] When he cited Psalm 19:1 in support of his program, Bacon was only reflecting what was commonly assumed by his readers.[5] He united arguments for science with a conviction that natural knowledge would ultimately benefit humanity and true religion. Nowhere is this more poignantly displayed than in a prayer embedded in his *Great Instauration:*

> Therefore do Thou, O Father, who gavest the visible light as the first fruits of creation, and didst breathe into the face of man the intellectual light as the crown and consummation thereof, guard and protect this work, which coming from Thy goodness returneth to Thy glory. Thou when Thou turnedst to look upon the works which Thy hands had made, saw that all was very good and didst rest from Thy labors. But man, when he looked upon the work which his hands had made, saw that all was vanity and vexation of spirit, and could find no rest therein. Wherefore if we labor in Thy works with the sweat of our brows, Thou wilt make us partakers of Thy vision and Thy sabbath. Humbly we pray that this mind may be steadfast in us, and that through these our hands, and the hands of others to whom Thou shalt give the same spirit, Thou wilt vouchsafe to endow the human family with new mercies.[6]

To Bacon, steeped in biblical modes of thought, inquiry into nature could not be completely divorced from faith, since the Christian heritage demanded above all love of God and neighbor. In his unfinished utopia, *New Atlantis* (1627), Bacon weaved medical, natural, and religious knowledge into a seamless garment that showed the benefits accruing to a society where love of God was united with perfected knowledge. Yet Bacon's interpenetrating concerns of natural philosophy, humanity, and religion were but a reflection of widely held beliefs among early moderns. Astronomers, natural philosophers, and theologians all acknowledged the necessity of divine assistance and the gift of divine light to understand the heavens.

The shared assumptions of early moderns were many, and the programs they offered for understanding nature were diverse. In this chapter, I sketch the different strategies by which scholars read the Book of Nature and the Book of Scripture. Such diversity constitutes the intellectual backdrop against which debates about Copernicanism and the Bible take on significance. The Book of Nature was built by analogy on the Book of Scripture and became a metaphor so ubiquitous that it appeared in treatises as diverse as Galileo's *Il Saggiatore* and Renaissance works on judicial astrology. Such diversity raised deeper questions of how the two books related to one another. Did both teach the same underlying principles or truths? Or did the content of each book contain different truths to be discovered independently of one another? Did the Book of Nature function only to confirm the truths of the Book of Scripture so that science could be used as an apologetic for religion? What was required of an interpreter when one book seemed to conflict with the other?

COSMOLOGICAL DIVERSITY IN EARLY MODERN EUROPE

One of the most important problems confronting sixteenth-century astronomers was realism. The medieval encyclopedia held a division of labor between technical astronomy and natural philosophy, the former attempting to save the phenomena without any claim to represent the actual motions of the heavens, while the latter sought to explicate true causes behind the phenomena.[7] The early modern period witnessed an increasing tendency for astronomers to assert that theories of celestial motion represented a true picture of the heavens. This realism is undeniably found in Kepler's second major work, the *Astronomia Nova* (1609), but the move to take astronomical constructs as real representations began before Kepler. In the prior generation, many astronomers (e.g., Tycho Brahe, Christoph Rothmann, Caspar Peucer) offered modified Ptolemaic (geoheliocentric) systems that revealed their assumption of realism in astronomy. Even Ptolemy's defenders were engaged in an ongoing reevaluation of which sciences would give access to truth.[8] We are far from understanding the origins of early modern realism, but it seems likely that it resulted from a number of sources in the sixteenth century. I suggest here two sources that antedate *De Revolutionibus*.

One important source for the ascendancy of realism in the sixteenth century was the cosmographical tradition begun through the reintroduction of Ptolemy's *Geography* in the fifteenth century. Standard accounts of Copernicanism stress how modern realism grew out of the Copernicans' insistence on the physical reality of the heliocentric system. These accounts typically explain Kepler's strong physicalism as growing out of a deep commitment to Copernicus, but this commitment does not explain why Kepler chose the unusual title *Mysterium Cosmographicum* (1596) for his first book. The word *cosmographia* had found currency in works deriving from the Florentine revival of Ptolemy's *Geography*. Jacopo Angelo offered a complete translation in 1406 and self-consciously changed the title to *Cosmographia* to fit his Latin audience. Many editions appeared through the fifteenth and sixteenth centuries and inspired writers to produce works of their own on cosmography.[9]

The revival of Ptolemy's *Geography* did not seem to have much effect on astronomy in the fifteenth century, but the cosmographical tradition did require the use of some astronomy to perform its tasks. Ptolemy distinguished sharply between chorography (topography) and geography proper, the former treating specific locales on the earth in the immediate surroundings, and the latter the place of the earth from a global point of view. Geography required placing the earth in the universe as a whole. The most notable expansion of this tradition was the greater emphasis given to the use of mathematics in addressing geographical questions. Francesco Maurolico's *Cosmographia* (1543), cast in the form of a dialogue, gives prominence to the mathematician Nicomedes, who is pictured as the contemplator of the divine works. Nicomedes devotes himself to speculation or contemplation—the necessary attributes of the cosmographer. Anticipating Kepler's *a priori*

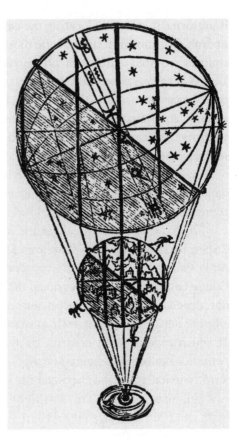

Diagram from Peter Apian's
Cosmographicus Liber (1524)
showing the position of the
earth against the celestial globe.
Courtesy of the Notre Dame
Library.

mode of argument, Nicomedes insisted that the sphericity of the earth could only be proven by reason since the senses fail us for this task.[10]

In the early sixteenth century, another cosmographical work appeared that was to pass through many editions and become enormously influential, Peter Apian's *Cosmographicus liber* (1524). Apian developed the Ptolemaic definition of cosmography in one significant way. He refined Ptolemy's categories by drawing a three-way distinction between cosmography, geography, and chorography. For Apian, chorography remained the description of particular locales, but geography described the world by referring to mountains, rivers, etc. that produced a picture of the relative positions of places on the earth. Cosmography treated the location of the earth in the universe as a whole.[11]

It performs this task by locating the earth against the celestial sphere. Apian's diagrams indicate how he thought that each of the three disciplines required a different observation point. Since cosmography looked at the earth within the larger setting of the celestial globe, its goals began to merge with those of astronomy and to require a consequent emphasis on mathematical techniques.[12] Although the two scientific traditions of astronomy and cosmography were somewhat separate, their underlying philosophical commitment to mathematical realism brought them together in the late sixteenth century. The cosmographical tradition did not offer a new cosmology, but it did become a crucial preparatory step.

The growing sense of realism also appears in a work that antedates Copernicus's by five years, Girolamo Fracastoro's *Homocentria* (1538). An Italian physician who was later appointed as chief physician for the Council of Trent, Fracastoro argued that a philosophy of nature should not depend too excessively on authorities but on experience. His empirical attitude shows itself in many of his medical treatises, one of which expounds the doctrine of sympathy (*de sympathia et antipathia*). Like other Renaissance thinkers, Fracastoro held that there were systematic relations between the superior (supralunar) and inferior (sublunar) worlds. This no doubt explains why he thought of his work as a contribution to the welfare of religion and the Christian republic. In his preface to *Homocentria* Fracastoro could congratulate Pope Paul III on the discovery of new celestial orbs because the realities of both worlds confirmed Paul's pontificate.[13]

Fracastoro argued for the higher dignity of astronomy than the discipline had received in Aristotle's classification because of its methods and the inherent dignity of its subject matter. The nobility of "those divine bodies" was seen to require the precision of mathematics, which for the most part proceeds from certainties (*ex certissimis*). The combination of mathematical precision and observational confirmation demonstrated astronomy's predictive success and argued for a higher consideration than it had normally been accorded. For Fracastoro, astronomy participates in a kind of divine knowledge (*cognitio divinitatis cuiusdam*). Because it yields a real theory of the physical heavens, he explicitly claims that astronomy leads to truth itself (*ad veritatem ipsam*) about the stars.[14] It appears that the realism of Kepler and other Copernicans has discernible roots, or at least precedents, in figures like Apian and Fracastoro.

A complete cosmology was also offered by the chemical philosophy, a version of the universe that paralleled and at times competed with other Renaissance traditions. Deriving mainly from the writings of Paracelsus (Theophrastus von Hohenheim), the chemical philosophy sought to reform both natural philosophy and medicine in a holistic program designed for the benefit of the human race. This union of medicine and cosmology was predicated on correspondences between the macrocosm (universe) and the microcosm (humans). Since processes in the macrocosm were essentially the same as those in the microcosm, the study of disease was thought possible by investigating the underlying laws of nature in both. The chemical philosophy focused heavily on empirical observations, and its rhetoric against standard modes of natural discourse often sound similar to the criticism leveled by Bacon and other empiricists.

No better representative of this philosophy can be found than the German Oswald Croll, whose *Basilica chymica* became one of the most influential chemical treatises of the seventeenth century.[15] Croll maintained, as did many Paracelsians, that true knowledge of nature depended on a prior knowledge of God. He rejected "the philosophical errors of the heathen" in favor of a genuinely Christian philosophy because only Christians have "the seed from God by means of regeneration which the heathen do not have."[16] How was this requisite regeneration to be judged? Croll insisted that it was not by a verbal profession of faith. As evidence, he praised Hermes Trismegistus as a pre-Christian holy man whose knowledge of the secrets of nature demonstrates his enlightenment by the Holy Spirit. Croll was not one to judge matters on surface appearances. Having received a doctor of medicine degree from Geneva in 1582, Croll could not have avoided the influence of Calvin's successor at the Genevan academy, Theodore Beza. Calvinists tended to be deeply suspicious of non-Christian philosophy while stressing the necessity of personal conversion. Regeneration, associated in ancient thought with baptism, became in Calvinism a more direct encounter with God, resulting in conversion. In this way, Croll united the pursuit of nature with religious enlightenment.[17] The study of the true chemist, according to Croll, involved a simultaneous

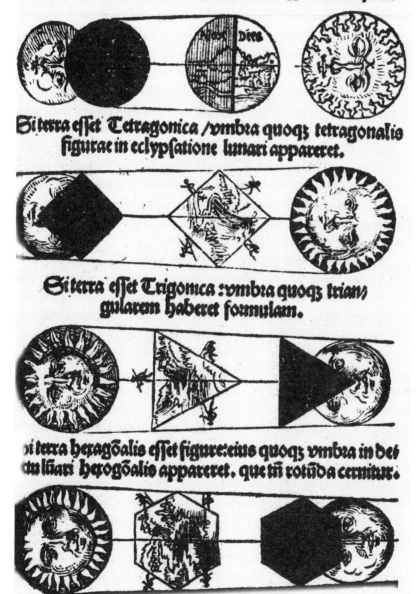

hoc Schema demonstrat terram esse Globosam.

Si terra esset Tetragonica /vmbra quoqȝ tetragonalis figurae in eclypsatione lunari appareret.

Si terra esset Trigonica: vmbra quoqȝ trian gularem haberet formulam.

oi terra hexagōalis esset figure: eius quoqȝ vmbra in de ctu lūari hexogōalis appareret. que tñ rotūda cernitur.

Diagrams from Peter Apian's *Cosmographicus Liber* (1524) illustrating the sphericity of the earth. Courtesy of the Notre Dame Library.

pursuit of "the frame of the world" and of oneself. The external world and the internal world were identical in essence so that no one without a knowledge of himself could have "any true intrinsic and essential knowledge of things."[18]

This reading of nature had profound implications for how the chemical philosophers read the Bible. Nature as God's acts and Scripture as God's word had principles that matched one another. And what could be more profound a mystery than the Trinity? The English Paracelsian, R. Bostocke, saw the *tria prima*—salt, sulphur, and mercury—as reflections of the Trinity, a connection that led him to conclude that only three major categories of disease correspond to the three principles.[19] Just how tightly the word of God could be tied to nature is evident in Croll's defense of "Physick" (i.e. the use of medicines). Answering those who refuse to take prepared medicines even in their serious need of them, Croll avers:

> We ought not to resist the ordinance of God. There is a twofold Physick: visible or created and Invisible, even the WORD of GOD. It is by the WORD of GOD therefore that any one whoever is restored to health. He that despises his WORD despises Physick and so on the contrary. For he that says Physick is worth nothing does upon the matter affirm there is no God.[20]

It is hard to imagine a more damning opprobrium for a seventeenth-century thinker than to be accused of atheism, but Croll did not shrink from equating the rejection of medicine with a judgment that would consign the recalcitrant to the flames of hell. I doubt Croll's condemnation is hyperbole. Following Paracelsian thought, Croll saw the word of God no less in the administrations of prepared medicine than in Scripture itself. Reading nature was a mirror of reading Scripture. The truth of the one was identical with the truth of the other. Although the chemical philosophy shared little with the cosmographical tradition and Fracastoro, it did represent the growing tendency to offer a comprehensive cosmology that united several different sciences.

Another tradition of reading the heavens *and* Scripture emerged in seventeenth-century England. We noted above how John Ray extolled the Creator through the diversity and precision of the creation. It was particularly in England that a new genre of literature arose in which the metaphor of God's two books played a central role: physico-theology. This literature was characterized by discussions of various features of the universe that were then related to attributes of God. Two manifestations of the new genre that appeared quite late, but that also reflected a trend extending well back into the seventeenth century, were William Derham's *Physico-theology* (1713) and *Astro-theology* (1715). While Derham allowed that the divine attributes and honor are seen no matter what system of the universe one adopted, he insisted that the "new system" was a better manifestation of the glory of God. The new system could not be completely equated with the Copernican because Copernicus had retained the sphere of the fixed stars. The new system was indeed heliocentric but infinite.[21] Copernicus, of course, knew nothing of Galileo's discoveries of the moons of Jupiter, but by the turn of the eighteenth century astronomers had discovered five Saturnian satellites. This suggested to Derham that there were worlds yet to be discovered, inhabitable worlds that existed in other solar systems, for Derham accepted an increasingly common idea of each fixed star being the center of another system.[22]

An infinite universe, far from detracting from the centrality of God, indicated for Derham the infinite power and wisdom of the Creator. Why would God have made an infinite universe? To show his glory to humans, who read his characters in the heavens:

> I shall only entreat all my readers to join with me in their earnest prayers, that as this work is designed for the good of mankind, particularly for the conviction of infidels, for the promotion of the fear and honor of God, and the cultivating of true religion, so it may have its desired effect.[23]

How did God construct the Book of Nature? Like so many before him, Derham took Psalm 19:1 as his cue. The most striking feature in the language of nature was its clarity. Even the heathen can deduce from the heavens that there is a god.[24] The clarity of the celestial witnesses is confirmed by the universality of religion. Derham takes Cicero's survey of theism in *De natura deorum* as confirmation of the Psalmist's words that no one is bereft of abundant evidence of the deity. By 1700, Europeans were much more aware of human diversity and culture than they were in 1500. This expanded knowledge had a profound effect on how Christian Europe framed its arguments for God and for the truth of the Christian religion. Yet the discoveries of the intervening two centuries only added to the conviction, for someone like Derham, that nature, when read consistently and carefully, confirmed the existence of God.

Derham's appeals to nature as arguments for the existence and glory of God went beyond a general level, a salient feature of the physico-theological literature. The distances between the planets in the solar system, particularly the earth from the sun, indicated how "convenient their situation" and how harmoniously this system had been fitted.[25] So, for Derham, God is not only the Creator but the contriver of an intricately constructed system, the details of which confirm the divine wisdom.[26] Derham argued for the prime mover to explain planetary motion in a fashion similar to Thomas Aquinas's adaptation of Aristotle. Derham was willing to admit a long series of natural causes going back to the sun, but the motion of the sun could only be explained by "the same Infinite Hand that at first gave them Being."[27] Derham is only one representative of a much wider phenomenon that began in the seventeenth century and continued until the nineteenth, the natural theology tradition. At the end of the early modern period, Derham could assume agreement between God's two books that made it possible to draw detailed comparisons between the two.[28] The literature of physico-theology represented a firm belief in the physical reality of the Copernican system. More significantly, it fostered the tradition of linking the Book of Nature with the Book of Scripture in very specific ways.

The early modern period witnessed the arrival of a diversity of cosmological schemes that competed with one another for hegemony, but all were characterized by an increasing sense of responsibility to delineate a true account of the heavens. This underlying assumption of realism, however, yielded no answers for how the heavens would be read nor what tools were best for reading them. Nor did it yield answers for how best to read the Scriptures with regard to natural truths. Historians of the scientific revolution once took it as a given that the tumultuous changes in cosmology in the early modern period resulted from Copernicus's groundbreaking work. While *De Revolutionibus* played a central role in the emerging realism of science, it did not play an exclusive one. A diversity of cosmological programs preceded and followed Copernican advocates. Nonetheless, all these new readers of the Book of Nature seemed to share the common goal of arriving at truth. As they pursued that goal, they also shared Copernicus's assumption of the benefits that cosmology would bring to religion and to humanity at large.

READING THE SCRIPTURES: CRISIS & CONTINUITY

If the early modern period saw great changes in how the heavens were read, it equally witnessed a thoroughgoing transformation in how the ancient Scriptures were interpreted. The dramatic proportions of the hermeneutical crisis occasioned by the Reformation can be gauged by the response of the Roman Catholic hierarchy at the Council of Trent. The fourth session of the council issued its decree concerning Scripture and its proper interpretation on 8 April 1546, a decree that defined the Catholic hermeneutical approach to a large degree during the next several hundred years. It influenced the proscription of the Copernican theory in 1616 and the later trial of Galileo in 1633.

The original decree of the council did not have in view any issues of natural philosophy; it was directed against the individualizing tendencies of the Protestant movement and served as an answer to the Protestant critique of the Roman Catholic doctrine of tradition. Three aspects of the decree are important for our purposes: the relation of Scripture to tradition (what is the source of dogma?), the Church as authoritative interpreter (who settles dogma?), and the patristic consensus on the meaning of Scripture (how is dogma determined?). All these had an important bearing on natural questions, a point that we shall scrutinize more carefully later.

On the Protestant side, the Reformation was no unified movement, though there were many attempts to mount a united front against Rome. The two most important movements, Lutheranism and Calvinism, had many differences that were accentuated as time went on. One indicator of how deep the differences could be shows itself in the suspicions many theologians had about Philipp Melanchthon because of his leanings toward Calvinism. Yet, on one crucial methodological doctrine, all Lutherans were united with Calvinists against Rome: the notion of *Sola Scriptura*. The terminology probably derives from Melanchthon himself, but the concept lay at the root of each Reformer's answer to the problem of hermeneutical authority.

Sola Scriptura was one in the litany of phrases that summarized the Protestants' objections to "popish religion" (*sola gratia, sola fide, solus Christus*). It meant that the canonical Scriptures were the sole source and criterion for determining doctrine. The church was not to be the hermeneutical authority; rather the Scriptures were to critique the church. When faced with the problem of competing interpretations, the Protestants invoked the ancient directive to compare Scripture with Scripture (*Scriptura Scripturae interpres*), i.e., a comparison of texts that would illumine each other. Such a heavy reliance on Scripture itself demanded corollary notions of the clarity and sufficency of the Bible. When the Protestants were confronted with the Catholic charge as to how they knew whether their interpretations of the Bible were correct, they often responded with an appeal to the clarity of the biblical text that made an authoritative interpreter superfluous. In the Catholic view, the bedrock of hermeneutical certainty lay in the church's reliance on the unanimous agreement of the church Fathers. In the Protestant view, while the patristic witness was venerable, it could not be the foundation of proper interpretation. The Bible alone held that place of honor. This foundational difference resulted in an interpretative crisis for Western Christendom, but most of the polemics did not bear directly on issues of natural philosophy. They had to do with more narrowly theological issues such as the real presence of Christ in the Eucharist, justification by faith, and the role of the papacy in the universal church.

All the differences separating Catholics and Protestants demanded a common base of argumentation, and theologians on both sides found that grounding in the *sensus litteralis*. Luther, for example, often chided the medievals for their excessive reliance on allegorical interpretation, and he often took recourse to the historical meaning of biblical texts in order to fend off Catholic criticisms. The Catholics too appealed to the *sensus litteralis* to counter the Protestant rejection of the doctrines of transubstantiation and the papacy. This emphasis on the *sensus litteralis* by both Catholics and Protestants came to have an important bearing on how different disputants construed the relation of Scripture and nature.

One indication of how hermeneutical differences affected the relation of Scripture to natural philosophy appears in a mid-seventeenth-century work of Catholic apologetics. In 1644, a book appeared in Pisa that combined the Tridentine requirement of patristic unanimity with an assessment of cosmological issues suggested by the text of Genesis and by the philosophical opinions of the ancient Fathers. Paganinus Gaudentius's *De errore sectariorum huius temporis Labyrintheo* consisted of three small works, two of which dealt directly with natural philosophy, *Conatus in Genesin* and *De philosophicis opinionibus veterum Ecclesiae Patrum*. Gaudentius's work posed the question whether the ancient church Fathers unanimously held certain philosophical positions regarding nature. A patently polemical treatise, *De errore sectariorum* sought to show that Lutherans, Calvinists, and other sectarians could not claim the support of Holy Scripture or the church Fathers. The Protestants argued that the Tridentine requirement of patristic unanimity was unworkable for two reasons. The ancient Fathers often held contradictory opinions among themselves, and they sometimes held opinions unanimously that are clearly false. By the Tridentine criterion, Catholics are bound to believe these opinions that conflicted with bona fide knowledge of nature. Gaudentius did not recoil from this thorny problem. For example, he clearly affirmed that the heavens are spherical while acknowledging that the church Fathers sometimes denied their sphericity. Their denial was not ludicrous, however, since the heavens do not appear completely spherical to the eye. Many of the Fathers quoted biblical words (Isaiah 40:22; Job 22:14) indicating that the heavens are more like a vault than a sphere. Gaudentius calls attention to Ecclesiastes 1:4ff. suggesting the heavens are round. In spite of these different indications in Scripture itself, many Fathers held that the heavens were neither mobile nor spherical. Was this patristic witness devastating to the Catholic hermeneutic of dependence on the Fathers? Gaudentius responded:

> In those matters that do not touch on the faith and which are not contained in the express words of Sacred Scripture, it is not necessary for us to follow the interpretation of Church writers, especially in those matters that are more physical than theological.[29]

Gaudentius, writing after the condemnation of Galileo, appeals to the same distinction Galileo made between matters of faith and matters of natural philosophy. Trent only committed Catholics to those things in which the Fathers agreed unanimously and which were matters of faith and morals (*de fide et moribus*). Thus, if the Fathers unanimously affirmed beliefs which were absurd, it was because such beliefs were not matters of faith and morals. Furthermore, on many issues of natural knowledge, the church Fathers had no unanimous consent.

Not all exegesis of the Bible in the sixteenth century was controlled by the dynamics of Catholic and Protestant polemics. In fact, different interpretations of those sections of the Bible touching most directly on natural philosophy cut across the ecclesiastical divide. Early modern commentary on Genesis did not appear *de novo*. Sixteenth-century interpreters were heirs of a long and highly diverse tradition of commentary on the hexaemeron, the six-day creation in the book of Genesis.[30] Several patterns of commentary are evident from the period when the church Fathers began elucidating the meaning of biblical texts. One pattern treated Genesis as giving detailed physical descriptions of the world from which cosmological inferences could be made. Another interpreted Genesis and other texts bearing on natural philosophy as having less direct correlations between language and reality. This latter approach saw biblical concepts such as the "floodgates of heaven" as metaphorical descriptions of supraphysical realities that were not directly visible to the naked eye of the observer. Such an approach did not deny historical or physical realities behind the text, but it did emphasize that those realities could not be mapped one-to-one onto the words of the biblical texts.

These patterns of hexaemeral comment were set even before the Edict of Milan (313) when Constantine legitimized Christianity in the Roman Empire, well before the appearance of most commentaries on Genesis. Though we know little about the Roman Christian Minucius Felix, his dialogue *Octavius*, modeled on Cicero's *De natura deorum*, pitted the pagan Caecilius against the Christian Octavius, who emerges victorious from their debate. Minucius employs a combination of Christian and pagan sources to develop a design argument for the existence of a supreme God.[31] The background of natural theology taken from

Roman and Greek sources allows Munucius to argue that a proper reading of the natural world in all its detailed variety would lead one to the same conclusion that the Scriptures teach: there is only one all-knowing Creator who providentially rules the course of nature. The details of motion, sound, light, and wind all testify to the *Summus Moderator* whom Christians worship.[32] This apologetic context of early Christian discussions of nature shaped both polemical and exegetical literature on Genesis in the early centuries of Christianity because the commentators always had to interact with pagan thought on nature.

Basil the Great (303–79), the Cappadocian Father, gave a series of homilies on the hexaemeron that influenced later commentators. Like so many of the Fathers, Basil both drew upon and countered various natural doctrines from Greek sources. In his *Address to Young Men* (*pros tous neous*) Basil expresses a typical attitude among early Christian authors regarding the value of pagan literature. The Christian had to subordinate Greek learning to the Scriptures and to the pursuit of salvation, but pagan learning could also be valuable in the matters of this world and in understanding the Scriptures themselves. If Moses could gain wisdom by being trained in the wisdom of Egypt, then his followers could surely learn wisdom from pagan sources, provided such knowledge conformed to Christian truth.[33]

Basil's main purpose in expounding the six days of creation was to contrast a Christian view of the world with ancient pagan views, especially Manicheism. To do so, he had to focus on the *sensus litteralis* and place the kind of allegorical interpretation he was so familiar with in the background.[34] Basil's treatment of philosophy was eclectic, drawing on a panoply of Greek sources (Plato, Aristotle, Stoics), but he also had an intimacy with later Hellenistic thought in Origen and Philo.[35] Basil showed two sides in his treatment of Greek natural philosophy. He adduced support for his interpretations when Greek philosophy did not contradict a received understanding of the Scriptures or Christian doctrine, but he also desired his hearers to move beyond a simple recognition of factual truth and to laud the Creator of all, a step that pagan philosophers were unable to take.

A second tack discernible in Basil's interpretations is his engagement with pagan ideas that conflict with Christian belief, the most prominent of which is the notion of eternal matter adopted by Plato and Aristotle. Basil argued, as his predecessors did and his successors would, that the world was made out of nothing by the one God. There are several corollary facets to this belief. Basil understood the order of the words in Genesis 1:1 as important, "In the beginning made God the heavens and the earth."[36] The Greek *arche* had the same systematic ambiguity as the Latin *principium;* both can be translated as "beginning" or "foundation." Since *arche* was used in Greek philosophical terminology, Basil took both meanings as essential. He saw the creation as the beginning of time and as the foundation of the world structure.[37] He found the pagan rejection of Genesis to be ironic in that the very *arche* that Greek philosophers sought was to be found in the God Christians worship. As the *arche* of the world, God created time and matter simultaneously:

> Now God, when he decided to start bringing into existence the non-existent things, (before any of the present visible things), he at once knew what kind of thing the world should be and generated together the suitable matter with its form. For heaven he set apart a nature that was fit for it. And for the form of the earth he laid down an essence (*ousia*) that is proper and necessary for it. He formed fire, water and air as he wished and brought them into being as the *logos* of each separate thing demanded.[38]

Basil unites several salient features of his theology in this short passage. He contends that God simultaneously fit matter (*hule*) and form (*eidos*) together in a prearranged manner, including the classical distinction implied between different natures for the celestial and the terrestrial worlds. Each natural phenomenon (fire, water, air) has its own *logos* (*ratio*). Here, Basil also affirms the dependent relation that the world sustains to God while maintaining that God has ordered material reality with a structure and rationale.

Basil's reading of Genesis defies facile classification. On the one hand, he takes the words of Genesis as referring to physical entities, not to mythico-poetic celebrations. On the other, his controlling context of interpretation demands that he separate Christian belief in creation from pagan philosophies. This latter orientation requires Basil not to constrict the meaning of the biblical text too tightly, as if one could read Genesis in a formulaic manner. This sublety of reading would suffice to protect his hearers from the enticements of pagan learning by commending what was positive in the philosophers while fending off what was harmful in their explanations.

Augustine's own circuitous path to Christianity prepared him for a type of exegesis that made encounter with natural philosophy inevitable.[39] Augustine's hermeneutical-theological method consisted of two important parameters. One was to show the unity of truth within the Scriptures themselves by reconciling texts that seem to be contradictory on the surface. His second was to show the unity of truth between the Scriptures and natural truths. He used natural knowledge available to him to provide a framework in which biblical concepts could be explicated.[40] This attempt to reconcile scriptural passages with natural knowledge necessitated the recognition of accommodation in the Scriptures, a feature that Origen had invoked already in the third century.

Although Basil's commentary had hints of accommodation, it was Augustine who most fully recognized that the text of Genesis was not a straightforward journalistic account of the six days.[41] He emphasized repeatedly that the language of Genesis was adapted to the weak understanding of humans, who could not pierce into the depths of a simultaneous creation.[42] Augustine held that the totality of the material world was created in an instant of time at the beginning, a concept that explained how the sequential account of Genesis could be reconciled with Ecclesiasticus 18:1, which taught a simultaneous creation (*creavit omnia simul*). Augustine found this latter text useful in guarding against two harmful tendencies, one of which made matter coeternal with God; the other portrayed God as a tiresome laborer who was so powerless that he had to create repeatedly over the course of six days. The first error came from Plato, Aristotle, and the Greeks more generally; the second derived from the Manicheans.[43] The latter criticized Genesis as a crude, irrational cosmogony with its representation of the deity as requiring six days to accomplish the task of creation.

Augustine's mode of reasoning influenced hexaemeral commentary for centuries. His interpretations did not so much positively assert what Genesis taught as they placed a boundary around faulty interpretations to be avoided. For Augustine, there is an underlying inscrutability of the created world, and all the interpreter of Genesis can do is offer plausible readings of the text. Most importantly, the interpreter must show what interpretations are *not* acceptable. Unlike those interpreters who essentially illustrated the meaning of Genesis from nature, Augustine posed problems for himself that arose out of a consideration of the text vis-à-vis reason and nature. Augustine saw problems and conundrums where others saw assertions. What was meant by the term "heaven" in verse 1? Was it physical or spiritual? How could Genesis say that God created light when no celestial luminaries were yet created? Was this an intellectual (nonmaterial) light or physical light? The words "God said" occur like a refrain throughout Genesis, chapter 1; was this spoken in time or in his eternal Word? Augustine's questions seem endless. In most cases, his answers are quite tentative, or his questions are left unanswered. Augustine's unparalleled influence in Western theology and hermeneutics created an environment for a great latitude in the interpretation of Genesis with regard to nature. The medievals continued Basil's and Augustine's approaches to Genesis by interacting with natural philosophy, in particular Aristotle's, in their interpretations of Genesis.[44] Early modern interpretation was a search for continuity with Christian antiquity because it was a period of hermeneutical crisis. Just as the diversity of cosmological schemes sought for roots in the natural philosophy of antiquity, so the existence of theological controversy demanded looking to the past to resolve the crisis of the present.

THE HEXAEMERAL TRADITION IN EARLY MODERN EUROPE

The number of hexaemeral commentaries grew dramatically in the Renaissance, no doubt due to the invention of printing in the fifteenth century as well as to the polemics of Catholic-Protestant debate in early modern Europe. Surprisingly, only recently has sustained attention been devoted to the hermeneutics of Reformation figures, even of the magisterial Reformers themselves. With Luther, we step into a different world of hexaemeral interpretation.[45] Although well-versed in the Augustinian tradition, Luther departed from Augustine's method in his own exegesis of Genesis. He often cited Nicholas of Lyra, who discussed issues of natural philosophy extensively in his hexaemeron, but Luther eschewed arcane philosophy in favor of a literal exposition that read the words of Genesis in an everyday sense.[46] He especially disdained allegorical readings because he thought they denied Moses' intention to teach about the visible world. He did not look with favor on previous hexaemeral commentary, nor on the ability of philosophy to adequately grasp the "grandeur of this subject matter."[47] It is not clear why Luther chose to reject his own Augustinian heritage in his approach to Genesis, but I offer here a tentative proposal based on some other salient facts of his life.

By the sixteenth century, biblical commentators were much more intent on expounding the literal, historical sense. While this was one option among four in the medieval scheme of hermeneutics, it became the predominant mode of explanation in early modern exegesis.[48] In one respect, such an emphasis might lead the interpreter to engage issues of natural philosophy, since the text was being read for its physical meaning. This did not occur with Luther. His reading of Genesis entailed physical referents for the words, but he was not constrained to interact with philosophical issues of nature. I suggest that Luther's denigration of philosophy originates in his rejection of Scholasticism. His lectures on Genesis began in 1535 but they reflect a sentiment expressed six years earlier, in 1529, at the Marburg Colloquy. One of the most important issues in the Reformation was the presence of Christ in the Eucharist. The ancient Christian tradition almost uniformly affirmed the real presence. Luther too affirmed it but he understood it differently than he did the doctrine of transubstantiation defined by the medieval councils (e.g. Lateran IV). Other Protestants such as Zwingli took Christ's words (*hoc est corpus meum*) symbolically only. At Marburg (1529), Luther and his cohorts attempted to come to terms with Zwingli over this crucial issue, but in the end they could not come to agreement. Luther took the words of consecration literally; it is the body of Christ. Yet when queried about the grounds of his position, Luther rejected philosophical explanations and relied instead on the "plain meaning of the words." Christ said it; the issue was settled. Luther as a man of the sixteenth century focused on literal, historical interpretation, but his break with Rome was not over a few dogmatic points; it entailed a rejection of philosophy that was associated with medieval dogmas.

The other magisterial Reformer, John Calvin, produced even more biblical exegesis than Martin Luther. I will return to Calvin's influence in the Netherlands in chapter 5; here I only wish to indicate some of the broad features of his treatment of Genesis.[49] Calvin was aware of ancient commentary on Genesis as well as medieval Jewish interpretation. He often invoked interpretations that found precedence in the earlier history of hermeneutics. Both Jewish and Christian interpretation had invoked the notion of accommodation in explaining the famous problem of Genesis 1:16: "and God made two great luminaries, the big luminary to govern the day and the small luminary to govern the night. He also made the stars." Calvin followed this tradition by explaining how Genesis could speak of the moon as one of the two great lights when it was well known that the other planets were physically larger than the moon. According to Calvin, Moses' description employed "common language" that did not require arcane knowledge of astronomy. Genesis spoke according to the appearances.

While Calvin followed previous hermeneutical tradition in this regard he did depart from Augustine's specific solution for the six days. Augustine had argued that the six days were a literary device to explain a simultaneous creation, thus accommodating human limitations, an explanation that aided his argument with the Manichees in their denigration of Genesis as having an inferior deity. Augustine insisted that the Almighty God did not need six days for his creative work; the six days were not a reflection of actual history but a literary accommodation. Calvin agreed that Genesis was an accommodation to human limitations, but it was not the text that was adjusted; rather, the actual history was adjusted to human understanding. God, in Calvin's interpretation, actually took six days to bring the world into its present state, a process that displayed the Creator's providence over time. I suggest that there are two reasons for Calvin's departure from Augustine's mode of explanation. Calvin did not face the Manichean challenge threatening Augustine, so a retreat to literary accommodation was unnecessary. Yet interpretation in the sixteenth century also reflected a shift from the concerns of the ancient church. The ancients tended to take it as given that allegorical interpretation was normal. One had to justify *ad litteram* interpretation that sought to understand the *sensus litteralis* or *sensus historicus*.[50] Calvin, on the other hand, assumed that the literal sense of Scripture was the normal mode of reading the text. By the sixteenth century, a greater historical consciousness had emerged in which reading ancient texts, including the Bible, required a first level of interpretation that sought to connect the structure and meaning of the text with the history it described. With regard to Christian interpreters, Augustine's voluminous work on Genesis may very well have been a leading cause of this turn to the historical sense, but it was certainly ensconced by Calvin's time.[51]

While the magisterial Reformers engaged natural philosophy only in an incidental fashion in their commentaries, other figures dealt with such problems more extensively. Protestant hexaemeral commentators differed widely in the extent of their knowledge of natural philosophy. One of the most important was the Italian convert to Protestantism, Jerome Zanchius. During his long and illustrious career as an exegete, Zanchius contended for his vigorous Calvinism with close attention to biblical texts. When he turned his attention to Genesis, Zanchius recognized what most all expositors acknowledged:

> There are two divine books in which God deemed to speak to inform us about his eternal essence and his perfect nature as well as about his best will and highest love toward us. The first is the Book of Creatures or Works, the other is that of Sacred Scripture or of the word of God. If you but compare these a little, you may see that though they are different, they are distinguished by this: to show forth or work together for this end, the knowledge of God and our happiness.[52]

Zanchius recognized that these two means of displaying the Creator's praise had different audiences, the Book of Nature being universal and the Book of Scripture being directed only to God's people Israel in the old covenant. Although the Book of Creatures was prior to the Book of Scripture, the latter is "more distinguished in perfection, clarity and dignity."[53] Probably due to his years in an Augustinian monastery, Zanchius was well grounded in Aristotelian natural philosophy, which he used extensively in his commentary on the hexaemeron. He treated a wide range of scientific topics in his *De operibus Dei*, writing in effect a "practical textbook" of Aristotelian science with the purpose of demonstrating the concord of Genesis with established knowledge of nature.[54] In Zanchius's mind, Genesis only outlines God's creative work, the details of which must be filled in by careful scrutiny of God's works. The Spanish Jesuit Benito Pereira followed a similar tack in his *Commentariorum et disputationum in Genesim* that appeared in four volumes in Rome from 1589 to 1598. A contemporary of Clavius in Rome, Pereira no doubt had access to a wide range of natural knowledge in the Society of Jesus, a knowledge that is displayed with great erudition throughout the commentary. Neither Pereira nor Zanchius, however, attempts to deal extensively with the changes in natural knowledge that were occurring in their day, a fact most probably due to their assumption that recent and unproven knowledge is inappropriate for incorporation into a biblical commentary.

A quite different tone appears in the commentary of Marin Mersenne, a figure of great importance for science in the seventeenth century. Mersenne, a prolific correspondent with major figures of early modern science, became a priest in the order of Minims in 1611. He acquainted himself with most of the scientific developments of the first half of the seventeenth century, employing them extensively in his attack on emergent atheism in France.[55] This polemic began in his first published work, *Quaestiones celeberrimae in Genesim* (1623), a work that took the Genesis text as an occasion to explain how the newest developments in science confirmed the Christian faith.[56] Mersenne was conscious of two separate sets of interlocutors. The first were those who claimed that Catholic theologians are tied to Aristotelian natural philosophy; the others were those who oppose the Catholic faith in the name of the new sciences. Mersenne's strongest attacks were reserved for the skeptics who denied any possibility of true knowledge, either of nature or of God. "In this volume," says Mersenne, "atheists and deists are confronted and dismissed, and the Vulgate version is justified against the calumny of the heretics."[57] His method drew upon whatever knowledge science would yield, and his commentary encompassed everything from music to optics. For example, when explaining the words of Genesis 1:3, "let there be light," Mersenne illustrated the nature and properties of light by his and others' work in optics.[58] Genesis gave truth about nature, but only science can detail the structure of nature through empirical inquiry.

Mersenne's apologetic against atheism and the role of his Genesis commentary in that polemic must be understood against the background of both Renaissance naturalism and the natural magic of the sixteenth century.[59] Mersenne's commentary, divided into three major parts, attacked the naturalist views of Julius Caesar Vanini in the first part. Vanini followed closely the views of the Paduan Pietro Pomponazzi, both of whom excluded any supernatural intervention in the course of nature. Consequently, this naturalist line of thinking led Pomponazzi and Vanini to explain away many Christian miracles as natural phenomena employing little-understood devices. According to Mersenne, this neo-Aristotelian naturalism made too great a division between faith and reason. Reason shows, for example, that humans have a free will, and so the astrological determinism of Pomponazzi and Vanini violate this basic premise. Similarly, reason demonstrates the existence of an immortal soul; such a tenet is not based only on faith. Mersenne's critique argued that the stars were not predictors of future events, thus rejecting the common practice of judicial astrology.

The last part (part three) of *Quaestiones celeberrimae in Genesim* attacked the other extreme, Renaissance magic. If Pomponazzi and Vanini had separated reason and faith too much, the magi confounded nature and spiritual forces too readily. Such a course would be a disservice to religion in the end. Mersenne's criticisms were directed against Georgio Veneti's *In Scripturam Sacram,* which attempted to explicate scriptural miracles by appeal to magical powers. Neo-Platonic thinkers like Pico della Mirandola and Marsilio Ficino had argued that the influence of the stars could be avoided by human knowledge of them and by strategies to circumvent their powers. Ironically, their insistence on human freedom assumed the astral influence so emphasized by the naturalists. The common correlation between certain metals and certain planets (e.g., iron with Mars) was arbitrary and spurious, according to Mersenne.[60] Mersenne walked the thin line between naturalism that excluded divine involvement in nature, and magic that attributed too many spiritual powers to nature. In this way, he signaled the middle road that Christian thinkers had to walk with regard to nature. According to Mersenne, the faults of these positions were related to how Genesis was read. Genesis should not be read in too naturalistic a fashion, nor as a kind of spiritual Kabbalah. Genesis, for Mersenne, showed the Creator's intimate involvement with natural laws that were sufficient to explain much in the daily course of the world.

Genesis was treated by the early moderns as a source of knowledge about the natural world for two reasons. They inherited the ancient Christian assumption of its divine origin, and therefore believed that it bore on natural knowledge in some way. They also were much more inclined to handle the text of Genesis according to the *sensus litteralis* (*historicus*), a change that seems partially attributable to the conflicts between Catholic and Protestant hermeneutics. This twofold context meant that the interpretation of Scripture with regard to nature would be impelled by continuity with earlier interpretation, and imperiled by conflicts over how acceptable interpretations would be determined.

Interpretations of nature and Scripture in the early modern era reflected a common assumption among their practitioners, namely, that there could be no real conflict between proper knowledge of either. This assumption was intimately linked to the belief that nature, religion, and society were mutually interdependent. Results in one sphere of inquiry should always reinforce the others. This goes far to explain how the early moderns could argue for an increasing sense of realism in the interpretation of both Scripture and nature. Astronomy in the sixteenth century made a transition from the task of saving the celestial appearances to that of giving a true account of their structure and nature. Biblical hermeneutics, too, moved increasingly away from a multifaceted approach to a concentration on the historical sense. Such an emphasis would inevitably mean that cosmologies involving a moving earth would have to face the hermeneutical challenge of reconciling terrestrial mobility with entrenched scriptural readings.

CHAPTER TWO

Copernicus, the Bible, and the Wittenberg Orbit

*T*he heliocentric theory was not new in 1543 when Copernicus's great work *De Revolutionibus* was finally published at Nuremberg. It had become known throughout learned Europe by word of mouth and by the circulation of Copernicus's manuscript *Commentariolus* that was finished in 1514. These new hypotheses received a largely positive welcome among astronomers, a reception that moved Georg Joachim Rheticus (1514–74), junior mathematician at Wittenberg, to cross national and ecclesiastical boundaries in order to study firsthand this revival of the Aristarchian theory.[1] In the summer of 1539, Rheticus traveled to Frauenberg and within a year he published his *Narratio Prima* (1540), the first public exposition of the heliocentric theory in modern times. That so seminal a book as *De Revolutionibus*, written by a Polish canon, was printed in Protestant Germany was only one of the ironies that fill the story of the reception of Copernicanism. This revolutionary work was not taken seriously as cosmology until almost fifty years after its publication. Proclaimed as a great simplification, Copernicus's system turned out to be as complicated as the Ptolemaic system it eventually replaced.[2] A theory ostensibly atheological raised fundamental hermeneutical questions and had profound theological implications.

The central figure in the reception of Copernicanism at Wittenberg was, surprisingly, not Martin Luther but Philipp Melanchthon. Luther's casual dismissal of the earth's motion in his *Table Talks* of 1539 is difficult to interpret and may not reflect the true thinking of the magisterial Reformer.[3] In any case, Luther's objections cannot be understood as informed opposition but may simply be antipathy toward novelty or toward audacious claims to explain the heavens. Little evidence exists that Luther's views had any philosophical or theological influence in the sixteenth century.[4] The matter is quite otherwise with Melanchthon. His opposition to the Copernican theory was more substantial and based on a wider range of knowledge and concerns. His citation of scriptural texts has long been noted, but a deep understanding of the hermeneutics behind these citations and their relation to natural philosophy remains to be achieved.[5]

In this chapter, I portray the differing responses to Copernicanism at Wittenberg in the 1540s by showing how those responses are related to the theology of Lutheran confessionalism. It is as impossible to understand the reception of Copernicanism in German lands apart from the theological confessionalization of Europe as it is to grasp the theology of Lutheranism apart from the sociopolitical context in which it was embedded.[6] The astronomical and theological developments in the last half of the sixteenth century provide the background for the differing cosmologies offered by Tycho Brahe, Christoph Rothmann, and Johannes Kepler. The Protestants of Germany clearly wanted to promote the restoration of astronomy and the renewal of natural philosophy, but there was disagreement on how this renewal should best be accomplished. Perhaps it is not surprising when we consider that a similar diversity of programs existed for the reform of the church. Although the differing tendencies of Luther and Melanchthon caused theological divergences and battles among the heirs of the Lutheran Reformation, it was Melanchthon's influence that shaped the discussions of natural inquiry in the end. We shall see that Melanchthon's advocacy of natural philosophy grew out of a concern to found the moral and civil order on legal structures implied in the natural order, an attitude that fostered astronomy and other sciences.

THEOLOGICAL METHODOLOGY IN EARLY LUTHERAN CONFESSIONALISM

Three developments in the first two decades of the Wittenberg reform are indispensable for understanding the Lutherans' response to

Copernicus: the nature of the theological task, the role of patristic authority, and the process of confessionalization. The controversies of the 1520s provide the backdrop for Luther's and Melanchthon's methodologies, one of which was Luther's differences with Erasmus of Rotterdam. Both advocated extensive reform within the church, but they differed over the goals and means of those reforms. While Erasmus's program centered on moral reform, especially among the clergy, Luther insisted on more extensive purging of pagan practices that had arisen in the medieval church. A salient dimension of their differences revolved around the nature of the theological task. Erasmus's descriptive method drew directly on the humanist task of philology and displayed a reticence to make clear theological statements. In contrast, Luther saw theology as affirmation that responds to the word of God in the Bible. Theology was not an analytical task analogous to science but an affirmation of truth resulting in confession. In his debate with Zwingli at Marburg (1529), for example, Luther insisted on the real presence of Christ's body in the Eucharist as the only acceptable meaning of the words of consecration. Theological analysis was not called for, rather acceptance of the words of Christ himself who said, "this is my body." It is better to affirm than to analyze.[7]

Melanchthon, the primary author of the *Augsburg Confession* (1530), also shaped Lutheran theological method as early as 1520 with his *Loci Communes,* a work intended as an instructional aid for lay Christians, and as a guide for Lutheranism in its formulation of biblical faith. The method of topics (*loci*) was also employed in his natural philosophy lectures, but its significance went beyond its immediate pedagogical purpose. The topical approach implied that the choice of topics was to be brought to the Bible rather than to have the categories in the Bible elucidated. Textual citation and quotation were used to confirm already existent interpretations or to deny contrary ones; rarely were there hermeneutical arguments that weighed different interpretations of texts. The primacy of the biblical text in both Luther's and Melanchthon's theologies required a spirit of affirmation more than of analysis.

A second decisive step in widening the rift between Rome and the Wittenberg theologians came with the debates over the role of the patristic tradition and the infallibility of the church. Both Luther and Melanchthon denied the infallibility of the church and of the ancient councils. In *On the Councils and the Church* (1539), Luther clearly set forth the notion that ecumenical councils, while trustworthy guides and pedagogical examples, could not and should not be considered infallible. And what was true of councils was even more true of the patristic witness in general. Luther mined the Fathers for guidance and support, for wisdom and for warning, but he read them with a critical eye, arguing that the Scriptures can overturn all the opinions of the Fathers in the same way that the Fathers themselves wanted their works to be judged by Scripture.[8] This implied for Luther that the church must be primarily defined as a *communio sanctorum*, i.e., the people of God gathered around the Word of God. Over against the Enthusiasts who stressed the primacy of the individual being led by the Spirit, Luther stressed the centrality of Word and Sacrament, that is, the external means of grace, means that were given to the people of God directly.[9] In this scheme, there was no room for a papacy that Luther denominated "an Institution of Satan" only six years later.[10] This unique authority ascribed to the Bible required some corollary hermeneutical principle that would guide how the Scriptures were to be interpreted. Here Luther argued for two corollary attributes of the Scriptures: sufficiency and clarity. If the historic councils of the church had failed at times, what would insure that the church would never be totally misled with respect to fundamental Christian doctrines? Luther argued that Scripture alone was the sufficient source and criterion for dogmatic formulation. If traditions in the church also agreed with the Scriptures, all the better, but traditions were not to be seen as the basis of doctrine. Yet how could Luther know how to interpret the Scriptures apart from the church? Here he invoked the clarity of Scripture. Even the unlearned could read the Scriptures and derive the basic meaning of salvation from them.

Melanchthon offered similar arguments on patristic authority. Written in the same year as Luther's treatise and provoked by the Leipzig disputations, Melanchthon's *De ecclesia et de authoritate verbi Dei* reviewed many of what he judged to be errors of the Fathers contrary to the Scriptures. His purpose was not to reject the Fathers categorically, because his humanist respect for antiquity would not allow him to renounce the ancient church entirely. He sought confirmation of scriptural exegeses and doctrines in their writings, but he clearly made patristic authority subordinate to antecedent interpretations of the Bible. The decisions of the ancient councils are authoritative, argued Melanchthon, not on account of the councils themselves but on account of the Word of God (*propter verbum Dei*).[11] He inverted the relation argued by the Catholics (that the meaning of Scripture should be judged by the church's councils) and judged the councils of the church by the meaning of Scripture. Melanchthon was not unaware of the objection that the ambiguous passages of Scripture needed the judgment of the church to counter heresy, and he recognized the need of the church to make such judgments. However, this need did not justify a wholesale acceptance of all decisions. He only hinted at an answer when he claimed that "the writings of the apostles" (*scripta Apostolorum*) must be followed before "the pronouncements of the church" (*sententias Ecclesiae*).[12] Perhaps here Melanchthon intended some principle of comparing Scripture with Scripture so that a pattern of apostolic teaching could be derived. In the end, he insisted on a critical function for Scripture over a trust in the judgments of the church.

The third and final development was the process of confessionalization, which proved to be at once rapid and irreversible. From the onset of Luther's work in 1517 up to 1530 it was not clear what direction the Reformation in German lands would take, but in that latter year Charles V, the Holy Roman Emperor, called together a diet at Augsburg in Bavaria. This convocation—ostensibly an attempt to unify the disparate reform movements within the empire—unwittingly placed its stamp on the Lutheran churches as *confessing* churches, a mark that would forever distinguish these churches from simple biblicism. The act of confession as a declaration of faith had a long and entrenched history in Christendom, but with the Lutherans such an act took on a new dimension and significance, one that would affect how they would interpret both the Bible and nature. More importantly, it set in motion a series of theological controversies that are also reflected in the debates over Copernican astronomy and natural philosophy. These controversies remained strident long after the first generation of Reformers passed from the scene, and they continued unabated until the decisive step in the Formula of Concord (1580). After 1530, the act of confessing on the part of Lutherans signaled their attempt to define the parameters of orthodoxy apart from the see of Rome and to unify themselves into a well-articulated order of belief. The basis of all confession was to be the Scriptures, and the act of confessing had several functions: defining belief, teaching people, and opposing falsehood. The Augsburg Confession acted as a subordinate norm to the Bible, or what later Lutheran theologians called a *norma normata*.[13] This process of confessionalization allowed a disparate number of social movements and groups to form a cohesive bond united by their doctrinal distinctiveness.

These three aspects of the Wittenberg reform created a climate marked by strident polemics where the Bible played a key role in theological debates. Its role was heightened by its being made the central confessional norm that stood against papal and patristic authority. The use of the Bible in the process of articulated confession influenced the way in which it was read with respect to natural philosophy by emphasizing a hermeneutics of intertextual comparison (*Scriptura Scripturae interpretans*). Before following this development, however, we need to understand the early interpretations of Copernicus that appeared around 1543.

OSIANDER AND THE EARLY RECEPTION OF *DE REVOLUTIONIBUS*

During the publication of *De Revolutionibus* Andreas Osiander (1498–1552) penned one of the shortest but most controversial pieces ever written in the history of science. His *Letter to the Reader,* anonymously attached to Copernicus's masterpiece, has been decried as a betrayal of the Polish canon's intention and praised as a perceptive portrayal of the inherent limits of science. Ordained a priest at Nuremberg in 1520 and having introduced Luther's Reformation against the fanatics and Anabaptists, Osiander became involved in a series of theological controversies throughout his ministerial life.[14] During the years when he came into contact with Rheticus and the text of *De Revolutionibus,* Osiander was engaged in reforming the principality of Pfalz-Neuburg (1542–43).[15] Some twenty years earlier (1526), the city of Nuremberg had established a new school and invited Melanchthon to become its rector. Inspired by Luther's vision of fostering the Protestant cause through education, the Nurembergers were committed to reviving and reforming the liberal arts. Most likely, Osiander's interest in seeing *De Revolutionibus* through publication came from his interest in educational reform for German principalities.

The religious significance of Osiander's *Letter To the Reader* has been variously assessed, but certainly the most influential interpretation derives from Andrew White's characterization of it as a "Groveling Preface" that viewed Osiander as attempting to dismiss the physical claims of Copernicus. In White's view, Osiander may have believed in the motion of the earth, but he succumbed to the pressure of theological dogmatism by dismissing the question altogether. This had the unfortunate effect of denying Copernicus's central claim of the earth's motion. Some basis for this reading dates back to the early moderns— to no less than Kepler himself who, based on letters in his possession, exposed the anonymous preface as not coming from Copernicus's pen. Kepler clearly resented Osiander for the Nuremberger's attempt to play down the physical claims of *De Revolutionibus*, claims that Kepler wanted above all to promote. Kepler offered a critique of Osiander's reading of *De Revolutionibus*, an interpretation that had no implications for cosmology or for a physical theory of the planets.

What was Osiander's view of Copernicus, and was it motivated by theological concerns, positively or negatively? The interpretation of the *Letter to the Reader* varies among modern scholars, but the outline of Osiander's argument is clear enough. It runs as follows: *De Revolutionibus* claims that the earth is moving in a threefold manner. Although Copernicus's readers would naturally be repelled by this audacious claim, they needed only to be reminded of the purpose of a mathematical work. It was not to adjudicate the physical question of the motion of the earth; it was only to predict the motions and positions of the planets. If the reader kept this qualification in mind, he would see that the author had fulfilled his obligation as a mathematician admirably. Mathematics does not aim at knowing the causes of celestial motions, but only at saving the appearances. Since mathematics bases its fundamentals on geometry and employs variant hypotheses to predict the celestial appearances, there is no principled way to determine which (set of) hypotheses are the true ones. The sole criterion for settling on one set of hypotheses rather than another lies in simplicity. The natural philosopher seeks rather to arrive at what is probably true of the world, but neither he nor the mathematician can arrive at certainty. Certainty only comes from divine revelation. So let the reader not be deceived into thinking that here he will find truth, for in mathematics such is neither intended nor achieved.

To understand properly the religious implications of this *Letter to the Reader*, we must appreciate its disciplinary and astronomical significance. The most common modern interpretation of Osiander derives from Pierre Duhem, who argued that Osiander simply reiterated the ancient tradition of absolving astronomy of any responsibility to offer true theories of the universe. In Duhem's reading, Osiander was not against earth's motion as such; he rightly recognized the limitations inherent in science for claiming truth.[16] Robert Westman reemphasized the disciplinary context by claiming that Osiander was not advocating an instrumentalist view of science in general but only of *astronomy*, which did not have access to truth. Truth was the province of natural philosophy. Copernicus's physical claims violated accepted disciplinary boundaries by his audacious argument for the sufficiency of mathematical knowledge. Westman's Osiander sought to preserve the harmony of the liberal arts by limiting the geometrical devices to their proper sphere. Osiander leaves the question of earth's motion open; he only asserts where the answer *cannot* be found.[17]

We also have relevant evidence from Osiander's letters that he embraced the disciplinary limitation on astronomy for both philosophical and theological reasons. On the same day in 1541 (20 April), Osiander wrote to Copernicus and Rheticus, suggesting a way to obviate the objections of Aristotelians and theologians:

> The Aristotelians and the theologians will be easily placated if they hear that various hypotheses can be used to explain the same apparent motion. They are not offered because they are certain but because the calculation of the apparent and composite motions can be done as conveniently as is possible.[18]

> Concerning hypotheses, I have always thought of them, not as articles of faith, but as a foundation for calculation so that, even if they are false, they carefully exhibit only the appearances of the motion. . . . In this way, you may pacify the Aristotelians and the theologians that you fear contradicting.[19]

These epistolary indications are reflected in Osiander's own words in the *Letter to the Reader*. After explaining that the astronomer seeks only a basis for calculation, Osiander cites cognitive simplicity as a criterion:

> But when various hypotheses are occasionally offered for one and the same motions (as in the sun's motion, eccentrics, and epicycles), the astronomer will seize on the one that is the easiest to grasp. Perhaps the philosopher will rather require what is probable, but neither of them can understand or teach anything as certain unless it has been divinely revealed to him.[20]

Osiander's public and private statements seem consistent. The astronomer treats calculation without regard to truth, the philosopher has a kind of probable truth, and both can have certainty only by divine revelation. The private letters, however, seem to indicate a concern on Osiander's part that *De Revolutionibus* will pass the judgment of philosophers and theologians only if the hypothetical nature of Copernicus's views is emphasized. Does this mean that Osiander, in fact, embraced Copernicus's physical system, and only invoked the disciplinary distinctions because he worried over its reception? Oberman and Hofmann argued against Hooykaas that Osiander was even more convinced of the compatibility of Scripture and the physical interpretation of Copernicus than Rheticus was. They invoked a common distinction between Osiander's own conviction and his mode of presentation.[21] Oberman's and Hofmann's Osiander appears very different from Westman's. An answer may lie in the phrase *nisi divinitus illi revelatum fuerit* (unless it is divinely revealed to him). If Osiander equated "divinely revealed" with Holy Scripture, then he would be placing certainty in the theological arena, in which case he failed to deliver a judgment on the meaning of Scripture. If he thought, as Rheticus did, that the Copernican theory was physically true and that there could be no objection to terrestrial motion from the Bible, it seems likely that he would have taken the opportunity to exonerate the Copernican theory.

Why then did Osiander bypass the hermeneutical issue and resort to a traditional disciplinary distinction that absolved astronomy from any possible contradiction with truth? If Osiander believed in the physical truth of the Copernican system, as implied by White, Oberman, and Hofmann, it seems odd that he did not express himself more forthrightly. We know from the controversy surrounding him at Königsberg in 1549 that he did not hesitate to voice his disagreement over theological issues. He argued against most other Lutherans that justification was not so forensic (legal) as Luther and others had maintained. He stood virtually alone in his position and was publicly condemned.[22] Such a forthright man would hardly recoil from expressing his belief in the physical reality of the Copernican theory for fear of theological repercussions, an issue that was far less central to the Reformation cause than justification by faith alone.

Westman's Osiander seems much more consistent and believable. The undeniable disciplinary component looms large once we place Osiander in the larger context of early modern astronomy, but this contextualization only partially answers why he chose this traditional strategy. Osiander's work at Nuremberg suggests that pedagogical concern for the academy there involved reforming tendencies that sought balance between maintaining the proper order of the liberal arts and remaining open to innovation. After reading *De Revolutionibus*, he did not want such a useful tool to be dismissed by his fellow Germans who were engaged in establishing liberal arts in various schools. To smooth the way for the acceptance of Copernicus's work, Osiander invoked a traditional distinction that limited the truth claims of astronomy and thereby assured a reading of Copernicus. I suggest that he set the issue of biblical interpretation aside altogether because he focused his attention on the educational program at Nuremberg and hoped that the new astronomy would be an aid in the pedagogical task. Setting the hermeneutical issue aside was not something everyone was willing to do, but separating mathematical astronomy and cosmology did allow the study of Copernicus to proceed in an environment without constant theological wrangling.

PHILIPP MELANCHTHON: *THE PRAECEPTOR GERMANIAE* AND HIS NATURAL PHILOSOPHY

By the time Rheticus had returned from his sojourn with his Polish mentor, Melanchthon had long been the leading figure in education

Examen Eorum,
QVI AVDIVN-
TVR ANTE RITVM PV,
BLICÆ. ORDINATIONIS. QVA
commendatur eis ministerium
EVANGELII.
TRADITVM VVITEBERGAE
Anno *1554* Per
Philippum Melanthonem

15 · · 98.

Cum Gratia & Privilegio Electoralis Saxoniæ.

VVITEBERGAE,
Excudebat Zacharias Lehman.

Philipp Melanchthon as portrayed on the title page of *Examen eorum . . .*
anno 1554 (1598). Reproduced from the original held by the Notre
Dame Library, Department of Special Collections.

among German universities. His own involvement in establishing a school at Nuremberg in 1526, for example, was only one of many reasons why he became widely known as the *Praeceptor Germaniae* already by 1540.[23] Melanchthon sought to unite learning (*eruditio*) and piety (*pietas*) into a seamless garment that would clothe the German people with the necessary raiment to lead them back to Christ. This goal did not subordinate the arts to theology or mix their methods, but it did insist on the liberal arts (*litterae*) and theology laboring alongside one another to produce pious men who were at home in both pagan and Christian literature of antiquity. From the earliest days at Wittenberg his own career embodied this goal.[24] Although he was a member of the arts faculty, his first lectures included an exposition of Paul's *Letter to Titus* as well as classical authors. After coming under the influence of Luther, Melanchthon began to see many parallels between church and academy, and he began to believe that the reform of the former was linked to the establishment of the latter. His view of the desolate state of the liberal arts no doubt prepared him to hear Luther's message about the deplorable state of the church. His inaugural address as professor of Greek (1518) at Wittenberg sounded the theme of the restoration of languages for the advancement of knowledge and piety.[25]

Melanchthon's educational ideals reflected deeply his humanist training that had first been instilled in him by his uncle Johannes Reuchlin and by his contacts with Erasmus of Rotterdam. Reuchlin was the leading Hebraist of the German Reformation, and he taught his nephew to value the ancients and their wisdom.[26] Young Philipp was anything but slow in adapting to the rigors of classical languages and adopting the humanist vision of reviving Greek sources for contemporary consumption. He also seemed inclined to the golden mean, no matter what the topic. His pacifist nature, for instance, was later to frustrate more radical Protestants as he often attempted to formulate theological positions that maintained a *via media*. All this explains Melanchthon, for he was nothing if not balanced in his approach to learning and theology. He always sought to consider the wider implications of any particular formulation so that the special point under discussion could be seen in light of the whole. This predilection for balance was evident at the colloquy of Marburg (1529) where he attempted to unify the forces of the evangelical movement by urging

Luther to come to terms with Ulrich Zwingli over the issue of Christ's presence in the Eucharist. This orientation led him to seek the integration of disciplines and truth claims coming from any direction.[27] This catholicity of mind and integration of knowledge became one of the most important factors in his promotion of a plan for educational reform.

Natural philosophy had its own proper role to play in the reform of education in Melanchthon's mind. He has long been recognized for his promotion of mathematics and a wide range of natural sciences at Wittenberg.[28] In his preface to Sacrobosco's *De sphaera* (1531) Melanchthon eulogized astronomy as an indispensable aid to confound the atheists, and he discovered in Genesis 1:14 the most fundamental of all functions, the legitimation of the pursuit of astronomical knowledge.[29] Recently, his natural philosophy has been interpreted in the context of a theology of providence that sought to read God's ways and will from the natural order.[30] Melanchthon, following a long established Christian tradition, read all of nature as signs of God's ways with man and as an avenue leading to knowledge of the Creator. His concept of natural law played a crucial role in his natural philosophy, but it is also connected with the moral law. In his *Loci Communes* (1st ed. 1520) Melanchthon introduced a twofold sense of natural law in the context of his discussion of the Decalogue. The first sense of natural law contained what is normally understood in the special sciences (knowledge of numbers, order, logical syllogism, the principles of geometry and physics); these are definite and unchangeable. The second sense relates to principles of action "such as the natural difference between things which are honorable and those which are base."[31] Both mathematical-physical law and natural-moral law are in fact knowledge of "the divine law which has been grafted into the nature of man" and "implanted in the minds of men" although the second type is usually hidden from our recognition because of human sin.[32] Each type of natural law reinforced the other and connected the concerns of the natural philosopher with the concerns of the church and society at large. This becomes evident in the manner in which natural philosophy and Scripture interact in his *Initia Doctrinae Physicae, dictata in Academia Witebergensi,* lectures on physical science that Melanchthon delivered in 1549.[33]

The *Initia* treats a wide range of topics, showing a knowledgeable acquaintance with all natural philosophy of antiquity and a special intimacy with Aristotle. Yet Melanchthon's choice of topics and method of treatment can hardly be called Aristotelian. Since ca. 1531 he used the Greek term *physice* to denote the study of nature in general while he reserved *physiologia* for that pure knowledge of nature inherited from the Greeks.[34] Thus, when he begins the *Initia* with the question "*Quid est physica doctrina?*" the content of the adjective *physica* encompasses all of nature and its relation to God, but man remains the center of physical inquiry.[35] To divorce the study of heavenly bodies from the study of man would be to claim that they were created for no purpose (*frustra condita esse*) and to break the unity of natural laws. Such laws inevitably lead back to the Creator by a chain of causes explaining what unites and differentiates various aspects of nature.[36] The emphasis on causal inquiry is explained by Melanchthon's virtual equating of God's will (*voluntas Dei*) with the order of nature (*ordo naturae*) with respect to the natural world.[37]

If the marks of divinity can be read clearly in nature, they are equally evident on the pages of Scripture, and so Melanchthon conjoined causal inquiry into nature with the testimonies of Scripture in his pursuit of wisdom (*sapientia*). However, his use of the Bible in physical inquiries was no mere juxtaposition of two complementary entities; he attempted rather a causal inquiry into the meaning of scriptural texts themselves. As in all interpretation, Melanchthon sought to know how the words of the text related to the reader, a task requiring causal analysis. In his celebrated *Commentary on Romans* (1540) Melanchthon distinguished Christian patience from that virtue of the philosophers under the same name. Christian patience was not simply a moderation of grief but a positive acceptance of God's will which had three causes: the command to be patient, the knowledge that afflictions are not signs of wrath, and the promise of help. As with many biblical concepts, this was "beyond the grasp of reason."[38] Melanchthon's attempt to penetrate the meaning of biblical words by causal analysis affords insight into his approach to phenomena, whether of a natural or textual character. Knowledge comes through linkage of events and/or reasons behind the phenomena, reasons that exist independently of the inquirer. Natural inquiry looks to the things in themselves and seeks to understand the relations between particulars. Exegesis seeks to explicate texts by a series of links that are causally connected. Causal physical knowledge also plays an indispensable role in the church because it is linked to something that concerns both physics and the church—human life. Inquiry into the multiplicity of causes serves the church by discussing "many varied human actions." While there can be no confusion between physical theory (*doctrina*) and theology (*doctrina Evangelii*), "there are nevertheless many things in the church's doctrine which cannot be explained without physics."[39]

If physics is relevant to the church, it is equally true that the church's book, the Bible, is relevant to physics. In the section of the *Initia* under the rubric *De mundo* Melanchthon treated the subordinate question, "*Quis est motus mundi?*"—his answer calling on both physical and biblical reasons for rejecting the earth's motion. That the Bible could be invoked to oppose geokineticism was not at all obvious to Melanchthon's readers, "Although others ridicule a physicist [*physicus*] who cites divine testimonies, yet we think it only honest that philosophy be united to heavenly dictates and to consult, wherever we can, divine authority in the obscurity of the human mind."[40] Melanchthon's admonitions earlier in his treatise on quiet acceptance of the truth were reiterated in connection with terrestrial motion, perhaps because this topic was so strongly debated after Rheticus's return from Frauenberg. Luther's lieutenant was unambiguous about what the sacred text taught on this matter. "Let us be content with this clear [*perspicuus*] testimony" says Melanchthon, since the Psalmist taught the movement of the sun (Psalm 19:4–6) and the stability of the earth (Psalm 93:1).[41] As Melanchthon read the text, the Psalmist not only taught it, he "clearly affirmed" it. Similarly, he exhorted his student hearers to be content with "this clear testimony" of God's word. These readings of the biblical text, however, must not be understood as naive literalism, for they rested on a deeper hermeneutical issue that came to prominence in the 1540s. The Roman Catholics had affirmed the hermeneutical authority of the church in the fourth session of the Council of Trent (1546), in contrast to the Protestant claim that the Scriptures were sufficiently clear to discover the truths of salvation. Melanchthon stressed the clarity of the Bible on this and other divisive issues because biblical perspicuity was the foundation of his hermeneutics.

This is seen even more clearly in an important parallel between Melanchthon's physical and scriptural arguments that aids in understanding his scriptural opposition. Melanchthon unambiguously affirmed the certitude that could be gained in physical questions based on three criteria: first (foundational) principles, universal experience, and knowledge of consequences. From these three, a syllogism can be constructed in which the principle forms the major premiss while experience provides the minor premiss, the conclusion (*consequentia*) being assured by the logical character of the syllogism. But the truth of the premisses depends on other matters. The general principles that form the major premiss come from arguments drawn from physical phenomena (e.g., the circular motion of the celestial bodies). The truth of the minor premiss that depends on experience is assured by direct and indisputable observations.[42] Melanchthon's recurring phrases (*oculi sunt testes, oculi testantur, ab oculorum testimonio ducitur*) show him accepting empirical evidence at face value without analysis or question. This form of argumentation affords Melanchthon certitude when treating abstruse questions. Twice in the *Initia* Melanchthon addresses the question of the infinity of the heavens.[43] His syllogism runs thus:

No infinite body moves circularly (major),
Heaven moves circularly every twenty-four hours (minor),
Therefore, heaven is not infinite.

His major and minor premisses proceeded inexorably from indisputable observations, but the principle involved in the major premiss is constructed from a less direct form of experience. With mathematical astronomy providing the background, Melanchthon argued the universe must be spherical because of stellar motion around the poles, the rising and setting of the stars, and solar motion. All these, which indicate circular motion also imply a spherical universe.

This form of reasoning parallels his argument from biblical texts. Melanchthon must have been aware of the accommodation argument that biblical language was phenomenal and not to be taken physically, but he clearly understood the texts as giving physical descriptions. How could he be so confident of his interpretation? To explain this, we must have recourse to the Lutheran understanding of theology as affirmation (and confession) mentioned earlier and the generally Protestant doctrine of the perspicuity of Scripture. We may reconstruct a syllogism that describes Melanchthon's method of interpretation:

All Scripture is true and clear (major).
Scripture says that the earth does not move (minor).
Therefore, the earth does not (cannot) move.

No one denied the truth of the minor premiss. However, Rheticus, as we shall see, made a distinction between what Scripture says and what it actually teaches. Melanchthon took the direct experience (reading) of the relevant texts to assure himself that when the biblical authors say the earth does not move, they intended precisely what their words appear to say. The major premiss on the truth and clarity of Scripture was more complex, but Melanchthon considered the principle sure—like the physical argument about the circularity of the heavens—because all Christian theologies (Catholic or Protestant) considered the Bible to be true. Further, the clarity of Scripture was assured not by the Church (as with the Catholics) but by a careful weighting of different texts to arrive at a general principle. Just as various types of astronomical and physical knowledge yielded the principle that no infinite body moves circularly, so the comparison of various biblical texts yielded the principle that Scripture was clear. With this firm principle, Melanchthon was certain of a static earth. His physical conclusion rested on his belief that the finite cannot comprehend the infinite (*finitum non capax infiniti est*) so that an infinite world would throw nature and science into complete confusion.[44] So also, his hermeneutical conclusion rested on the belief that God would not deceive our senses when Scripture is read. To admit the inability to read Scripture or nature properly would have undermined the very possibility of natural and theological knowledge.

Melanchthon was keenly aware of those who would denounce him for invoking biblical authority to adjudicate astronomical or physical questions. There may have been two distinct parties he was countering. Scholastic thinkers, who kept questions of philosophy and theology quite separate, would not have precluded an agreement between the conclusions of natural philosophy and theology, but they would have strongly argued that the different types of argumentation cannot be confused. Melanchthon's willingness to use an amalgam of physical and biblical arguments may represent a humanist tendency to invoke historical-theological reasons when they seemed to confirm natural reasoning. The second group countering Melanchthon's approach might have been some humanists themselves whose approach to the interpretation of historical documents differed from his own. Although most Christian humanists of Melanchthon's age would have full confidence in the integrity of Scripture, some may not have had as great a confidence in the human ability to interpret it properly. This required Melanchthon to invoke a prominent notion in Reformation debates, the perspicuity of Scripture.

Melanchthon's opposition to terrestrial motion rested partially on Scripture, but he also had a deep distrust of novelty. By 1546, he had developed a much stronger historical attitude toward knowledge and the necessity of reforming the church. His educational program at Wittenberg and his own view of the inevitability of the Reformation were rooted in a growing sense of historical continuity. Four days after Luther's death on 18 February 1546 Melanchthon's funeral oration located Luther in an illustrious heritage of reformers stretching back to Adam, and his biography of Luther four months later gave a historical interpretation of Luther's life that resounded with a note of inevitability. Although Melanchthon argued against the Roman Catholic insistence on the authority of the patristic tradition, he used that same tradition extensively to counter charges of novelty leveled against the Reformation doctrines. This stress on continuity with the past and his condemnation of novelty in the *Initia* reflects this belief in the necessity of historical continuity to legitimate any proposed reform, whether ecclesiastical, theological, or pedagogical.[45]

From a modern nontheological standpoint Melanchthon's invocation of Scripture to condemn the motion of the earth appears naive and quaint—a mindless literalism that accepts the Bible as teaching a definitive theory of the world—but such an evaluation fails to appreciate both the details of his reasoning and the historical context of humanism that so influenced him. His advocacy of the liberal arts, especially mathematics, made him one of the leading figures for the spread of Copernican astronomy in northern Europe, and his own work in natural philosophy spurred others to pursue the same. In this light, his opposition to Copernican cosmology, specifically a moving earth, can only be explained by his requirement that physical truths must be derived from physical arguments because mathematical astronomy only saved the appearances at best. Since compelling physical arguments for the earth's motion were lacking in the 1540s, Melanchthon no doubt felt his physical conclusions confirmed by Scripture. He had no reason to give credence to the accommodation argument because he considered the Bible clear on the matter, since its various texts uniformly stated the immobility of the earth. The clarity of biblical texts served as the foundation of all his theological arguments because the admission of gross unclarity in the Bible would have meant capitulation to Roman hermeneutics and the death knell to the cause of the Reformation.

Rheticus's Realist Interpretation and Theological Defense

For Rheticus, the subject of the earth's motion in Scripture was not so straightforward. Melanchthon never identified his interlocutors in the *Initia*, but it is reasonable to think that Rheticus's *Narratio Prima* and Melanchthon's discussions of the physical truth of the Copernican system lay behind his scriptural arguments against the earth's motion. Rheticus no doubt felt the need to respond to the kinds of arguments that must have floated around the Wittenberg faculty, and because he had varied interests and wide learning, it comes as no surprise that he addressed himself to the question of the compatibility of the Copernican system with the Scriptures.[46] Our knowledge of Rheticus's work on the interpretation of Scripture comes from a letter of Bishop Tiedemann Giese to Rheticus dated 26 July 1543 in which Giese referred to a treatise where Rheticus reconciles the new view of the universe with Scripture:

I want your little work added where you have aptly vindicated the motion of the earth from disagreement with the Holy Scriptures. In this way you will complete the greatness of that well-grounded volume [*De Revolutionibus*] and will compensate for what is

CANON
DOCTRINAE
TRIANGVLORVM.

NVNC PRIMVM A GEOR
GIO IOACHIMO RHETICO, IN LVCEM
EDITVS, CVM PRIVILEGIO IMPERIALI,
Ne quis hæc intra decennium, quacunq; forma
ac compositione, edere, neue sibi uendicare
aut operibus suis inserere ausit.

LIPSIAE
EX OFFICINA VVOLPHGAN
GI GVNTERI.

ANNO
M. D. LI.

77

Title page of Georg Joachim Rheticus's *Canon Doctrinae Triangulorum*
(1551). Courtesy of the Notre Dame Library.

disagreeable where your teacher [*praeceptor tuus*] omitted mentioning you in the preface of the work. I interpret this as happening not from purposeful neglect of you but from apathy or carelessness and especially because he was becoming weak, for I am not ignorant of how he customarily made use of your service and willingness to assist him.[47]

Hooykaas claimed to have discovered this document to which Giese refers because the author of the *Treatise* refers to "my teacher" (*praeceptor meus*), a personal designation for Copernicus used by Rheticus in his *Narratio Prima*.[48] Further evidence lies in similarities of phraseology between Rheticus's *Narratio Prima* and those found in the newly discovered *Treatise*.[49] The tone of the document makes it impossible to discern whether it was written by a Catholic or a Protestant, and so Hooykaas argued that the work must have been written before or during the Council of Trent (fourth session, 1546) because a Catholic after the council would likely have referred to its hermeneutical decisions in his arguments. Since Johannes Campensis's *Enchiridion Psalmorum* is the latest book quoted, the document must have been written after 1532. This evidence correlates what we know about Rheticus's sojourn with Copernicus and his return to Wittenberg. Given the evidence of Giese's letter, it is probable that the *Treatise* was written between 1541 and 1543.

Making sense of Rheticus's hermeneutics is a more difficult matter because he often employs several different ways of reading the biblical texts. Since Hooykaas's *Commentary* has been the only extensive interpretation of the *Treatise*, I shall examine it critically, noting its fundamental flaws.[50] On the surface, Rheticus argued in a manner very similar to Galileo's by claiming a distinction between physical questions and theological truth, a form of argument that Hooykaas saw as directed against the literalists and as limiting the authority of the Bible to historical and religious matters, excluding natural questions from its purview.[51] The tone of Rheticus's *Treatise* stands in contradiction to the charge of novelty, a charge that was especially painful to those trained in the humanist tradition. Consequently, Rheticus labored to establish his continuity with St. Augustine by arguing that his own interpretations were consistent with this important historical precedent. This method of justifying interpretations based on historical precedent is also reflected in Rheticus's view of his teacher. In the *Narratio Prima* Rheticus sought to absolve his teacher from the charge of novelty:

> Concerning my learned teacher I should like you to hold the opinion and be fully convinced that for him there is nothing better or more important than walking in the footsteps of Ptolemy and following, as Ptolemy did, the ancients and those who were much earlier than himself. However, when he became aware that the phenomena, which control the astronomer, and mathematics compelled him to make certain assumptions even against his wishes, it was enough, he thought, if he aimed his arrows by the same method to the same target as Ptolemy, even though he employed a bow and arrows of far different type of material from Ptolemy's. At this point we should recall the saying "Free in mind must be he who desires to have understanding." But my teacher especially abhors what is alien to the mind of any honest man, particularly to a philosophic nature; for he is far from thinking that he should rashly depart, in a lust for novelty, from the sound opinions of the ancient philosophers, except for good reasons and when the facts themselves coerce him. Such is his time of life, such his seriousness of character and distinction in learning, such, in short, his loftiness of spirit and greatness of mind that no such thought can take hold of him.[52]

To Rheticus, departure from the opinions of the ancients was justified only when compelling empirical evidence to the contrary was available or when it was required by mathematical adjustments, a view he reinforced by his argument that Ptolemy had been understood correctly only by Copernicus and that therefore his teacher was in line with the ancient astronomer.[53] Similarly, in the *Treatise*, he argued that the Copernican theory could be absolved from apparent contradiction with the Scriptures on the basis of the greatest Latin Father of the church, Augustine. Answering the charge of novelty did not hinder Rheticus from endorsing a physical interpretation of heliocentrism, a belief requiring theological objections to be overcome if it was to find acceptance. From his earliest teaching days, Rheticus showed strong tendencies to search for causal explanations, concluding his inaugural academic address with Vergil's words: "Happy is he who can understand the causes of all things" (*felix qui potuit rerum cognoscere causas*). This was to become a theme of his work throughout his life.[54] The *Treatise on Holy Scripture* was of course superfluous if Rheticus had held to an Osianderian instrumentalist interpretation, and the text of the *Treatise* makes it clear that its author believed the new theory to be a true description of the universe.

The structure of the *Treatise* moves from general principles of interpretation to dealing with specific texts of the Bible. The historical foundation of Rheticus's interpretation derives from the Augustinian principle that research into the natural order is encouraged by God but must be done within the bounds of the catholic (i.e. universal) faith.[55] Rheticus concentrated on explaining the proper relationship between knowledge from the liberal arts and the teaching of Scripture since some had argued that the statements of Scripture have the force of demonstrative proof and therefore settle a question of natural knowledge. On the other hand, Rheticus's view of the relationship of natural inquiry and the interpretation of Scripture is more subtle than a rigid separation between science and religion. If Rheticus had held to such a strict separation, he would not have argued as he did in his *Treatise*. He distinguished at length between an article of faith taught in the Bible, which must be believed, and a physical description used by philosophers, which cannot be expected to be found in the Bible.[56] This distinction was not based, however, on a prior judgment of natural vs. theological issues but on natural questions that are settled because the catholic faith has defined them vs. natural descriptions that are written "in the manner of the philosophers."[57] He uses the doctrine of the creation as his extended example of a settled natural question. Even if Aristotelian philosophers could bring forth arguments in favor of the eternity of the world, for example, their arguments must be rejected as invalid because the question of the eternity of the world has been clearly answered by Scripture. The certainty of the doctrinal correctness of creation can be drawn from Scripture because in Holy Writ one can find a rationale for its teaching. God, through Moses, intended to lead the recipients away from idolatry and to the knowledge of the true God. That same God revealed his power and paternal care through the doctrine of creation.[58] Rejecting the notion of a created order would be tantamount to rejecting those theological teachings about God, and therefore it is proper to side with Scripture against Aristotle. At the same time, Rheticus claimed that the Holy Spirit, the divine agent of inspiration, did not intend descriptions of natural phenomena to be comparable to those offered by astronomers and natural philosophers.[59] The doctrine of creation does not fall into the category of descriptions that may be amended; it has apodictic (demonstrative) force because it transcends any particular natural question normally raised by philosophers. The motion of the earth, however, is an example of those issues that the divine Spirit appears to have avoided.

This method of interpretation does not correspond to an oversimplified natural vs. theological dichotomy but is based on judgments about the divine intention in Scripture. Hooykaas misconstrued Rheticus's argument by claiming that Rheticus (and Augustine) distinguished between those portions of Scripture that have apodictic force and those that contain time-bound ways of speech, the latter containing "no scientific statements in Scripture."[60] On the surface, Hooykaas's apodictic vs. nonapodictic distinction appears to find support in Rheticus's words:

> There was no better means of humbling the minds of such people and restraining their temerity than by withdrawing the authority enjoyed by other Scriptures in those texts of Scripture treating natural things (to the extent that a knowledge of nature is sought) and it openly teaches in this manner of treatment that the proper way to proceed is not by affirmation but by inquiry.[61]

This passage suggested to Hooykaas a distinction between passages in Scripture treating natural things and those treating historical and religious matters, the latter being fully authoritative while the former are not.[62] I suggest, however, that Rheticus's distinction is more subtle than Hooykaas recognized; it is not a means of placing various texts into authoritative versus nonauthoritative categories. The distinction has rather to do with levels of generality that are related directly to levels of authority. There are no other indications of authoritative versus nonauthoritative portions of Scripture in Rheticus's *Treatise* because for him Scripture without distinction has the force of a demonstration (*vim demonstrationis*), i.e., it is all authoritative.[63] When Rheticus appealed to Scripture to prove the doctrine of creation against Aristotelian philosophy, he necessarily included language in the Bible that contained references to nature. Thus, Rheticus's distinction cannot be cast in terms of theological (apodictic) portions vs. natural-scientific (nonapodictic) portions of the Bible. When he said that Scripture does not speak in the "manner of the philosophers" or give a description of nature, he meant that it would be improper to place the Bible on the level of a natural-scientific treatise and compare their relative merits. The Bible did not intend to give low-level descriptions of nature in this manner, but it also clearly spoke of nature.

How can one know when to take the language of nature in the Bible apodictically? Rheticus's answer is contained in two principles he outlined at the beginning of his *Treatise*. In the case of creation, a careful comparison of biblical texts reveals the theological relevance of Scripture. The natural language in these texts serves the deeper theological intention of revealing God. Apparently, for Rheticus no similar intention can be discerned with regard to the motion of the earth. His second principle deals with not going beyond the bounds of the catholic faith. That catholic faith had established long ago the doctrine of creation and its concomitant, the temporality of the world, and so it would be impossible for one who affirmed biblical authority to deny one of its central tenets based on a supposed distinction between natural vs. theological questions. Rheticus opts for a more general level of reasoning by comparing texts with texts and by being guided by historic doctrinal formulations. His interpretation relies on a distinction between low-level descriptions (which are not expected in the Bible) and higher-level affirmations that are culled from many texts and from historical precedent. The higher-level affirmations are properly subjects of theology while the low-level descriptions are colored by the common modes of speech employed by the biblical writers. These modes of speech should not be confused with physical descriptions because knowledge of nature should proceed by inquiry, not affirmation (*non affirmando, sed quaerendo tractandum est*).[64] Thus, Rheticus distinguished not between authoritative and nonauthoritative texts but between a combination of texts with a discernible theological intention and a specific text whose language is not directly theological but which employs physical language for a theological purpose. The interpreter must discern the divine intention behind a text, not whether the text is authoritative. The emphasis discussed earlier on confession (affirmation) in the early Lutheran Reformation can be heard behind these words; making the motion of the earth a theological matter would have effected a total closure on inquiry and that would have violated the Augustinian principle of the redemptive purpose of the Bible. On the other hand, if inquiry into nature should not proceed by biblical exegesis, it is equally true that understanding biblical texts should not be sought outside the texts of Scripture itself, a conclusion that leads Rheticus directly to specific texts.[65]

Reframing Rheticus's distinction yields understanding of his treatment of particular texts. When he turned from the general principles of interpretation to the specific question of the earth's motion, he appears to contradict himself. In his principles he argued that philosophical matters were not to be expected in Scripture and that one should not seek such principles since Scripture is not a philosophical textbook (*philosophicum librum*).[66] Yet in his treatment of motion, he argued *from scriptural texts* that the heavens must be immobile because they are called a *firmament*. He concluded that since the heavens must be motionless, motion must be attributed to the earth to account for the apparent motions of the heavens. Rheticus appears to have violated his own distinction by his contention that scriptural texts imply the motion of the earth. Hooykaas explained Rheticus's inconsistency as a rhetorical ploy; the rules of interpretation he outlined were applicable only to keep his opponents from quoting Scripture. He concluded that Rheticus "rather inconsistently" admitted the use of the Bible to corroborate the daily motion of the earth.[67] The portion of Rheticus's *Treatise* treating the biblical texts on the motion of the earth (end of section 16 through section 27) is very complicated and does appear to contradict his earlier espoused principles of interpretation. However, such a conclusion is not necessary if the specific language of Rheticus's text is attended to. First, his argument does not say that the motion of the earth can definitely be found in the Bible. He argues that the biblical language about the stability of the heavens (Isa. 40:22; Ps. 103:2,3) is allowed by St. Augustine and is consistent with the observed phenomena, noting that some texts seem to imply the motion of the earth (Job 9:6). The movement of declination that accounts for the seasons may be implied in David's Psalms (e.g., Ps. 73 [74 A.V.]: 17).

All these citations would imply that Rheticus thinks the motion of the earth is taught in the Bible in contradiction to his announced principle that natural-scientific teachings are not in it. Yet the key to resolving this apparent inconsistency lies in the last sentence of section 27: "These are perhaps the texts of Scripture which, if the earth is moving, can be said to contain something about the issue although obscurely." (*Hi sunt fere loci Scripturae, quibus si terra movetur, aliquid hac de re, sed obscure in bibliis possemus dicere contineri.*) The tone of this concluding sentence is by no means definitive and claims only that the motion of the earth, demonstrated on other grounds, is obscurely referred to in the Bible. This is not inconsistent with his claim that the Bible is not a book of natural philosophy. His argument is not contradictory if we read it thus: the Bible does not have a description of nature, as one finds in philosophy, but it does refer to the earth's motions in its various texts. Its references to the earth's motions are obscure, and we could never infer a theory of terrestrial motion from the Bible itself. But once we have a demonstration of the earth's motion from philosophy, we may look back into Scripture and see vague references to it. These texts, then, that refer to the motion of the earth cannot be taken as authoritative statements of dogma but they do obliquely refer to physical realities. Such an argument would have been consistent with the Protestant notion of the perspicuity of Scripture because this notion was never used to deny interpretative conundrums or textual obscurities. Rheticus subtly drew on this Protestant distinction to argue that the message of salvation was indeed clearly taught in the Bible but also that other, less central, truths might lie hidden under the redemptive message of Scripture.

The texts that mention the stability of the earth cannot be taken as referring to physical realities, according to Rheticus, or his entire argument would be vitiated. How then could such texts as Psalm 103 (104):5—"who founded [*fundasti*] the earth on its foundations, it will not be shaken forever"—be explained? To argue his case, Rheticus adduced texts that also speak of the moon and stars as founded (Ps. 8:4, *fundasti*), and yet it is clear that on the Ptolemaic theory these are never taken to mean that the moon or other planets do not move. Similarly,

Rheticus argued, the use of "founded" with reference to the earth does not necessarily imply that it is not moving. Rather, "founded" means that the heavens and the earth have been established in their regular course and that they cannot be changed except by the will of God.[68] The term "founded" may or may not refer to physical realities, but there is no compelling reason to take these texts as physical descriptions. To absolve the Copernican hypothesis from contradicting Scripture completely, Rheticus also had to treat a second group of texts in a nonphysical manner, those on the motion of the sun. After adducing those texts that appear to support a moving sun, Rheticus distinguished real from apparent motion. Here he fell back on his announced principle of accommodation and argued that one should expect the Bible to focus on apparent solar motion rather than real motion since the Scriptures repeatedly speak from the point of view of the everyday observer.

Rheticus's arguments were not motivated by any need to reinterpret the Scriptures in a manner that was inconsistent with theological tradition. His interpretations are both natural to the sense of the text and faithful to the history of interpretation. Indeed it would be difficult and dangerous for a sixteenth-century biblical interpreter to introduce interpretations that could not be supported by the Fathers of the church even though the degree of patristic authority was treated differently by Protestants and Catholics.[69] This suggests that Rheticus's theological resources were rich and varied enough to explain how the hypothesis of his teacher (*praeceptor meus*) and the Scriptures were not in conflict. Rheticus did not have to resort to nonliteral interpretations that were novel. Rather, he resorts to an ancient tradition of interpretation that recognizes the harmony of knowledge gained from empirical investigation and biblical statements.

If we cast hermeneutical methods in a literal vs. figurative dichotomy, we can only conclude that Rheticus was inconsistent in his handling of different texts. In some respects, his treatment can be characterized as nonliteral, but in other texts he applies a method that sees physical counterparts to the biblical words. Either Rheticus is being inconsistent with his announced principle that physical descriptions (*descriptio naturae*) are not in the Bible, or his method of interpretation is more nuanced than a distinction between literalist and nonliteralist would suggest. By *descriptio naturae*, I suggest, Rheticus meant that the

Bible was not attempting to give a theoretical account of the heavens but also that this did not mean that there were not physical descriptions in the Bible from the standpoint of a naive observer. The burden of his argument was to circumscribe the area of physical science (*physica*) so that physical questions would not be considered as settled simply by quoting Scripture. But the Scriptures did have some bearing on higher-level issues such as the eternity of the world. How could Rheticus distinguish between those questions on which Scripture had some bearing from those on which it did not? His answer was found in the ancient witness of the church. The church had long before decided that the doctrine of creation was *ex nihilo*, on which basis he could pronounce Aristotle's doctrine impious and blasphemous. However, such a conclusion did not imply that the Scriptures should be conclusive in every question of natural philosophy.[70]

THEOLOGY, ASTRONOMY, AND NATURAL PHILOSOPHY FROM AUGSBURG TO THE CONCORD OF 1580

While discussions about the compatibility of Copernicus and Scripture were taking place in the 1540s, the leading mathematician at Wittenberg was working on extensive revisions to *De Revolutionibus* to produce tables (ephemerides) for locating the positions of planets. Less than a decade after Copernicus's masterpiece was published, Erasmus Reinhold offered the *Prutenic Tables* (1551) as an extension and correction of the new hypotheses. Reinhold viewed himself as promoting Copernicus's work, something evident from his 1542 commentary on Peurbach's *Theoricae novae planetarum*:

> I know of a scientist [Copernicus] who is exceptionally skillful. He has raised a lively expectancy in everybody. One hopes that he will restore astronomy . . . I hope that this astronomer, whose genius all posterity will rightly admire, will at long last come to us from Prussia.[71]

Reinhold's *Tables* represent painstaking labor and considerable advancement over the tables contained in *De Revolutionbus*. For example, Copernicus calculated positions for every third degree and to the nearest minute of arc; Reinhold performs the same for every degree and to seconds of arc. In general, his positions for particular planets are much more accurate than those of Copernicus, which partially explains the extensive use of Reinhold's *Tables* in the late sixteenth century.[72] However, Reinhold's advocacy of Copernicus's work should not be taken as an endorsement of his cosmology. Since the pioneering study of Birkenmajer, Reinhold has been viewed in the framework of the "split interpretation" of Copernicus that Gingerich and Westman inter alia developed.[73]

The split between mathematical astronomy and cosmology with regard to Copernicus was extended throughout German universities, but perhaps the most prominent and influential figure in this development was Michael Maestlin (1550–1631) of Tübingen.[74] Maestlin's influence on Johannes Kepler was enormous, as he provided the kind of support that the younger scholar needed to venture his first and somewhat strange book, *Mysterium Cosmographicum* (1596), but Maestlin's own work and views reflected the division between astronomy and cosmology that characterized Reinhold. In his 1597 *Epitome Astronomiae* Maestlin held the traditional distinction between *theorica* and *sphaerica*, the former treating planetary motions without any concern for physical causes because the latter did so. It would be Maestlin's more famous student who would be the first to introduce physical considerations systematically into planetary theory. Yet Maestlin's background and training suggest that Kepler's concern was not wholly unprecedented. As we will observe in chapter 4, Kepler's choice of the term *cosmographicum* drew on a long tradition that reached back over a century, but it is quite likely that this tradition was transmitted to Kepler through Maestlin. Maestlin wrote at least one work on geographical problems, and appendices in his *Epitome Astronomiae* treat such problems as the measurement of distances on the surface of the earth, climatic zones, etc.[75] Maestlin's teacher, Philip Apian, used his father Peter's textbook *Cosmographicus liber* in his lectures while Maestlin was a student at Tübingen. Thus, although Maestlin seems to have kept issues of mathematical astronomy and physical causes quite separate, his own work on geography and his use of Apian's work may certainly have provided some grist for Kepler's physical speculations. In any case, by the 1580s astronomers in the Wittenberg orbit (e.g., Peucer,

Rothmann, Tycho) began to focus on the larger questions of physical cosmology rather than limit themselves to mathematical prediction.

On the theological side, the half century between the Augsburg Confession (1530) and the Formula of Concord (1580) proved to be highly significant for Lutheranism because several points of doctrine were debated extensively in a fashion that indirectly related to issues of natural philosophy.[76] Naturally, a major player in these developments was Melanchthon, whose teaching and writing spawned a whole host of Philippists with strong Calvinist tendencies. The many-sided controversies between Lutheran purists (Gnesiolutherans) and the Philippists included heated debate over the function of the Law (the ten commandments) in the life of the Christian, debates that unveiled the Philippist tendencies to focus on theological anthropology. In contrast to the Gnesiolutherans—who contended for the purity of the Gospel by downplaying any role for the Law in the Christian's life—Melanchthon's followers stressed that ecclesiastical and social cohesion required a formative role for the Law in the Christian life. Consonant with the educational and humanistic goals discussed earlier, Melanchthon wanted to provide the foundation of a thoroughgoing reform and so thought it necessary to go beyond the concerns that immediately faced the early Luther. Philippist theologians could be found throughout German-speaking areas and Scandinavia, and the 1560s witnessed the growing influence of Melanchthon's son-in-law, Casper Peucer, among the Wittenberg faculty. Peucer's views on the doctrine of the Lord's Supper reflected a cautious attempt to avoid controversy while embracing Calvinist views of the supper. Through Niels Hemmingsen (1513–1600) this Melanchthonian humanism had some influence at the University of Copenhagen, although his Calvinist sympathies brought his dismissal only five years after Tycho's famous lecture on the mathematical disciplines.

Many of the controversies during these fifty years revolved around the place and function of the law in Christian life. By denying that works had any part in salvation, Luther's theology stressed the role of faith in receiving divine grace. Yet Luther was no anarchist, and he sometimes spoke of works as important for the Christian and civil order but never in a manner that would lead back to the "Roman slavery" of justification by works. Salvation was by faith alone. After

Luther's death in 1546, the Gnesiolutherans and the Philippists seized on different aspects of Luther's teaching to emphasize now the Gospel, now the Law. Both parties held to the central polarity of Law and Gospel, believing that both were essential for balanced church life. Law without Gospel only condemned, while Gospel without Law might lead to libertinism. In the finer definitions of a later period (ca. 1545–80), this tension between Law and Gospel would prove to be one of the thorniest issues.

The majoristic controversy that grew out of the Leipzig Interim (1549) focused on the proposition "good works are necessary for salvation." Its major proponent, Georg Major, maintained justification by faith while also maintaining the necessity of good works. Nicholas von Amsdorf, Major's opponent, accused him of abandoning Luther's *sola fide*. The crux of the issue lay in what was understood under the term "salvation." For Major, salvation included not only initial justification but also the process of Christian living requiring good works. For Amsdorf and other Gnesiolutherans, salvation was more closely equated with justification that could not involve good works. To admit Major's thesis would be a wholesale capitulation to Rome, and the purity of the Gospel would be lost.[77] All this could be seen as a strictly theological debate except for its connection with moral and social issues. Melanchthon's followers and various governmental officials were concerned about maintaining social order; they thought it necessary to emphasize good works so that the Christian populace would not take Luther's doctrines as a pretext for immoral living and civil disobedience. In Melanchthon's mind, moral and civil order were preeminent and so the Law of the Old Covenant must play a positive role in the church and society. Morality and civility were based in turn on the natural order, and so natural philosophy played an essential role in undergirding the social structures of German lands.[78]

The analyses developed above suggest that the responses to Copernicanism in the orbit of Wittenberg differed according to how the relations of astronomy, natural philosophy, and theology were viewed. So long as participants like Osiander and Reinhold separated mathematical astronomy from physics and cosmology, they were able to encourage and pursue the use of Copernican planetary models and to develop more accurate ephemerides. This split interpretation reflected the medieval separation of natural philosophy and theology, so that the most conservative position available at the time was also the one that allowed the dissemination of the heliocentric theory. Osiander's animadversions, however, still reflect the ultimate (not temporal) priority of theology because biblical revelation was the only means toward certainty. Melanchthon, who appears on the surface to be the most conservative figure, emerges not as an opponent of astronomy but as a believer in the concurrence of astronomy, natural philosophy, and hermeneutics. While holding to a separation of astronomy and cosmology, he found the physics of Copernicus wanting for both physical and biblical reasons, a position that involved a degree of intermingling of natural philosophy and theology. Only Rheticus endorsed both the astronomy and the cosmology of *De Revolutionibus* and was thus faced with the necessity of defending heliocentrism from theological opprobrium. In these first decades, the proper relation of astronomy, physics, and theology was in flux, and the evaluation of Copernicanism required arguments over boundary disputes as much as internal assessment of its theoretical adequacy; nonetheless, neither form of evaluation determined the hermeneutical patterns used in scriptural arguments.

Both Melanchthon's and Rheticus's treatments reflect a methodological commitment to the physical truth of a world system, but their hermeneutics differed, albeit in ways more subtle than previously recognized. Neither disputant considered the Bible sufficient for deciding natural-philosophical issues, but both also thought it relevant to some degree. Melanchthon employed the Bible to confirm or deny conclusions arrived at through physical argument because he saw both physics and hermeneutics as concerned with human life and morality. His confidence in his interpretations depended on a belief in the clarity of scriptural texts analogous to his trust in the reliability of sense experience. Rheticus, on the other hand, thought references to natural things to lie deeper beneath the surface of biblical texts, though they were not totally excluded from the purview of Scripture. As with all later Copernicans, Rheticus appealed to accommodated language in the Bible, but he also mined the texts for positive arguments in defense of a moving earth.

Furthermore, Rheticus and Melanchthon shared a belief in the priority of natural inquiry over theological adjudication in answering natural questions, but they focused on different sciences for their prior certainty before resorting to hermeneutics—the former to astronomy, the latter to physics. Their exegeses both appealed to the patristic tradition—though Melanchthon's physical lectures did not do so explicitly—but they appealed to different aspects of that tradition. The Augustinian tradition invoked by Rheticus distinguished between descriptions characteristic of natural philosophy and theologically relevant doctrines, associating each type with distinct levels of authority. Melanchthon, in his hermeneutics, was not as subtle as Rheticus; rather, he stressed the clarity of a given text because he knew that the perspicuity of Scripture provided a foundation for Lutheran arguments against Roman Catholic interpretations. The exegetical differences between the two are based on their prior judgments about the divine intention in the composition of Scripture.

The intensity of theological debates in the last half of sixteenth-century Lutheranism raised the problem of how to interpret biblical doctrines properly, not only as they differed from Rome, but also as they fostered the internecine warfare of later Lutheran debate. These controversies came to a head in the concordist movement that produced the Formula of Concord (1580). This movement began in March 1577 with three theologians—Martin Chemnitz, Jacob Andreä, and Nicholas Selnecker—composing the Bergic Book that became the "Solid Declaration" in the Formula. The Formula represented the end point of one era and the inception of another by achieving a unity among Lutherans that had long been sought. After the Formula, Lutheran theology entered the relative calm of classic orthodoxy, a period characterized by a greater systematization of existing dogmas and a relative peace from internecine warfare.[79]

CHAPTER THREE

Geoheliocentrism and the Bible: Brahe, Peucer and Rothmann

*A*s the influence of Lutheran Wittenberg spread among other German cities and Scandinavia, many astronomers trained in the Wittenberg tradition began to take their places in various universities, where they taught both Ptolemaic and Copernican systems within a framework of geocentrism. Three of these astronomers (Tycho Brahe, Caspar Peucer, Christopher Rothmann) were destined to play important roles in the cosmological revisions of the 1580s and 1590s as more problems arose with theoretical formulations and empirical data. Of these astronomers, the name of Brahe alone became associated with a system that was immensely popular in the seventeenth century but that was eventually replaced with a modified Copernican system. By the last two decades of the sixteenth century, the cosmological implications of the search for an adequate astronomy were vigorously debated and the need for careful weighting of mathematical astronomy, a physics of the heavens, and the interpretation of Scripture became apparent.

Caspar Peucer, the son-in-law of the *Praeceptor Germaniae*, assumed a chair of mathematics at Wittenberg in 1554 shortly after the death of Erasmus Reinhold. Although he was to become primarily a physician, Peucer exercised an enormous influence in theological debates and had an active participation in the search for an acceptable cosmology. From the 1550s to the 1580s Peucer produced a number of works on astronomy, geodesy, and the astrological significance of the heavens. From his *De dimensione Terrae* (1550) to his *Hypotheses Astronomicae* (1571) over twenty years later, Peucer engaged in accommodating Ptolemaic astronomy to the insights of Copernicus.[1] His advanced lectures on astronomy regularly encouraged contact with Copernicus's *De Revolutionibus* but with the distinction between astronomy and cosmology that Osiander and Reinhold had maintained. It was not until the 1580s, after his extensive imprisonment for theological deviance, that Peucer corresponded with Brahe on matters of cosmological significance, although his interest in these matters is clearly evident in his works on divination.[2]

Christoph Rothmann occupied a position of growing importance as the court astronomer of the landgrave of Hesse, Wilhelm IV, at the court of Kassel where he and his patron pursued a regular plan of nocturnal observations. He arrived at Kassel in 1577 after studying theology and astronomy at Wittenberg, and he worked with the instrument-maker Jost Bürgi (1552–1632) to improve the landgrave's observations and star catalog. After his visit with Tycho Brahe on Hven in 1590, he appears to have returned to his native town of Bernburg in Anhalt, and to have been preoccupied with theological controversies.[3] Generally treated as one of the few committed to physical Copernicanism in the sixteenth century, Rothmann appears to have vacillated between heliocentrism and forms of geoheliocentrism, but he did have an important role in the debates over the celestial spheres in the 1580s. His thought on the relevance of the Bible to terrestrial motion remains completely unstudied.[4]

The Danish nobleman Tycho Brahe, whose castle on the island of Hven came to be one of the most prominent observatories in the sixteenth century, far exceeded Peucer and Rothmann in achievement and fame. His observational assault on Mars made an indispensable contribution to Kepler's elliptical paths and the area law. His goal, however, was not limited to correction of technical features of observational astronomy nor even to formulating the most adequate astronomical system. His correspondence shows him intensely interested in the total restoration of astronomy and the nature of the heavens. For him, neither Ptolemy nor Copernicus could lay claim to such a total restoration, the former because of mathematical absurdities and the latter because of physical absurdities. Tycho's task encompassed observational, mathe-

QVADRANS MVRALIS SIVE
TICHONICVS.

Tycho Brahe's Mural Quadrant from his *Astronomiae Instauratae Mechanica* (1598). Courtesy of the Notre Dame Library.

matical, and physical dimensions, a combination anticipating Kepler's explicit physicalism without an endorsement of Copernicus:

> I will rather labor to satisfy the heavenly appearances with our other hypotheses because, if I have the favor of the Author of heaven, I will work expressly for the restitution of the celestial motions

that the truth may be known. This will far exceed the Ptolemaic and the Copernican systems and rather correspond to the truth itself.[5]

The widely received view that the Tychonic system represented a compromise between an ancient Ptolemy and a modern Copernicus—an inevitable result of Tycho's commitment to an immobile earth—does not adequately reflect Tycho's own view of his task.[6] Recent studies have suggested that the origin of Tycho's geoheliocentric system may not be as inevitable as implied in the standard view.[7] Tycho repeatedly gave expression to an increasingly common goal among late sixteenth-century astronomers of searching not for adequate models of description that could save the phenomena but for the real system of the universe.

Tycho's project began with indisputable observations, because in his view all theoretical and cosmological consequences had to be consistent with more accurate data than were available in either Ptolemy or Copernicus. No doubt Reinhold's corrections to Copernicus had a decisive influence and suggested to him that even more accurate ephemerides were still to be produced.[8] The conclusions he drew from the new star of 1572, especially that it could be neither planet nor comet, also made him doubt that any system, ancient or modern, accurately enough captured planetary motions. But observations were only the beginning for Tycho. Observations could prove hypotheses wrong in physics but could not establish them. For example, celestial matter was a physical question, not an astronomical hypothesis, and no amount of accurate observations or mathematical elegance could affirmatively settle a physical question. The result of the cometary observations of 1577 and 1585 were for him purely negative—that the heavens could not have solid spheres—but they did not offer any positive information about the material of which the heavens were made. For a total cosmology one must also include physical arguments, chemical correspondences, and theological considerations as well.[9] Tycho summarized his global goal in a letter to Rothmann in 1588 with the following famous epigram:

You anxiously demanded to know what the depicted philosophical saying means: the inscription SUSPICIENDO DESPICIO (treating heavenly things) and the other DESPICIENDO SUSPICIO (treating earthly things). To accede to your wish, here you have it. You have rightly conjectured that this is a hieroglyphic for they have in view not only that superior celestial and that inferior terrestrial astronomy but also theology that is more divine but less commonly used. This is indeed a knowledge of all ethics i.e. the discernment of virtues and vices. It presents a physical consideration of created things.[10]

This view hearkens back to Melanchthon's vision of the interpenetrating concerns of physical inquiry and human morality, but it also adds a component with which Melanchthon was not familiar, a kind of Paracelsian unification of the celestial and the terrestrial realms by a series of correspondences, each of which revealed something distinctive of the whole.[11] Since Tycho explicitly emphasized theology as a category to be included in his program, it is reasonable to ask what role the interpretation of the Bible played.

Drawing mainly on Tycho's correspondence, I explore his interpretation of the Bible and what role his hermeneutics played in his emerging cosmology. I argue that his hermeneutics was far more subtle than the simple literalism typically ascribed to him and that it depended on his view of the relation of theology to disciplines such as astronomy and physics. One of his main correspondents, Christoph Rothmann, insisted that astronomers *alone* could rightly give the true system of the universe because astronomy alone offered *demonstrative* and *relevant* knowledge. Tycho argued that astronomy was limited to observational and mathematical prediction, while insisting on an observational difference between the Ptolemaic and Copernican systems.[12] His castle on Hven was established "for contemplation of philosophy, especially the stars," and it provided what he hoped would be a strong empirical basis for his restored astronomy, but it also housed a chemical laboratory in his cellar designed to study those correspondences between the terrestrial and the celestial realms. The physical structures around him reflected the inner structure of the great problem he posed for himself, that of revealing the true system of the universe. In Tycho's mind, this project had to include any and all methods of inquiry into nature that were available, including biblical interpretation.[13]

THE YOUNG TYCHO ON BIBLICAL INTERPRETATION

An important window on Tycho's approach to the Bible is offered by his inaugural address on the mathematical disciplines (*De Disciplinis Mathematicis Oratio*) which he delivered in 1574 at the request of the chancellor of the University of Copenhagen and the king of Denmark. This lecture, the commencement of a series on the fundamentals of astronomy, came as an invitation due to his fame for observing the new star of 1572 and the circulation of his *De Nova Stella*. While touching on a wide variety of subjects under the rubric "mathematical disciplines," Tycho focused on predictive astronomy and astrology, perhaps because he hoped to persuade the Danish king to support astronomical research since it would yield practical results in the political sphere.[14]

Tycho referred to the Bible throughout the lecture in connection with two major subjects: the history of astronomy and theological objections to astrology. In an overlooked portion of the *Oratio*, he presented an involved argument for the antiquity of astronomical inquiry based on biblical and Jewish sources. Tycho's audience knew that the ancient Pythagoreans were learned advocates of mathematics, but his story traced the discipline further back through the Egyptians and Abraham to the third son of Adam and Eve, Seth.[15] Nothing, of course, in the biblical narrative suggests that Seth was an astronomer; Tycho took his information from Josephus, the chronicler of Jewish antiquities. Relying uncritically on Josephus's history, Tycho argued that Seth's repository of astronomical knowledge passed on to Abraham who in turn transported it to Egypt during his sojourn there.[16] Like Seth, Abraham learned something of the Creator by observing the stars directly. Josephus and his coreligionist Philo were anxious to demonstrate that the admirable knowledge of the heavens so praised by the Greeks had its origins in fact in patriarchal sources, not Hellenic ones. Why Tycho should have been so anxious to treat the patriarchs as astronomers is less clear, but the answer may lie partly in his hermetic tendencies and partly in his argument for the theological acceptability of astrology.

Tycho's hermetic tendencies are reflected in his reference to Egyptian knowledge as well as in his claim that astronomy has Adamic origins. The hermetic tradition, revived and developed through the translation of *Corpus Hermeticum* by Marsilio Ficino in the late fifteenth century, had sought to legitimize its claims by appeal to an antiquity that antedated the Greek golden age. Many hermetic thinkers believed that the Egyptian Hermes Trismegistus had been a contemporary of Moses and was ultimately the source of Plato's thinking. Tycho put a new spin on this claim with his argument that the Egyptians had learned astronomy prior to Moses from Abraham, who was also in fact a recipient and conduit of secret knowledge having Adamic roots. To root astronomy in the original creation—in a prelapsarian human state—would not have been unusual for hermetic or Paracelsian thinkers. Most likely, Tycho drew this argument from his friend Peter Severinus whose *Idea medicinae philosophicae* had been published three years earlier and who offered one of the earliest systematizations of Paracelsian theory.[17]

As to theology, if Tycho could show that the contemplation of the heavens had its ultimate roots in the biblical patriarchs, this would certainly lessen any opposition that might be launched from a supposed biblical prohibition against astrology. Later in his lecture, Tycho took up the challenge of biblical prohibitions against secretive knowledge and practices.[18] He denounced false astrologers who engaged in mere fortune-telling and other more serious violations of biblical and astronomical knowledge. Yet he argued just as vigorously that the abuse of a science does not invalidate its legitimate use.[19] The ultimate basis of Tycho's argument came from Genesis 1:14–18, which taught that God placed the sun, moon, and other celestial bodies in the firmament as signs for times and seasons. The celestial bodies are God's servants (instruments) not only to show the harmony and beauty of God's works, but to indicate the hidden counsel of God to humans who read the heavens. Niels Hemmingsen, the leading Philippist theologian at Copenhagen, had argued for the freedom of God to perform his will against any limitation of divine freedom implied in an astrological connection.[20] Tycho agreed that God is "a perfectly free agent" but he was just as certain that God had chosen to employ secondary means to accomplish his will. The incarnation of the Son of God made it clear that God works through human instruments. Why then is it so strange that God would use the stars, as Genesis 1:14–16 asserts, to guide the crown of his creation, humanity?[21]

In addition to the Mosaic legislation, specific texts cited against astrology included Jeremiah 27:9, "But as for you, do not listen to your prophets, your diviners, your dreamers, your soothsayers or your sorcerers who speak to you saying, 'you will not serve the king of Babylon,'" and Isaiah 47:13, "Let now the astrologers, those who prophesy by the stars, those who predict by the new moon, stand up and save you from what will come upon you." Whether by prohibition (Jeremiah) or taunt (Isaiah), these texts seem to condemn any involvement with secretive or occult practices that could predict the future. How could Tycho defend astrology against this prohibition? Jeremiah's prohibition, argued Tycho, does not deny the signifying function of the stars but only exhorts the people of Israel not to fear the Babylonians, whose superstition and abuse of astrology are censured. Isaiah's taunt condemns "the Babylonians' vain confidence in their predictions"—not a proper use of astral influence for human benefit. Tycho appealed implicitly to the notion of the congruence of scriptural texts and to a *reductio ad absurdum* argument. Since there can be no contradiction within the Scriptures and since Genesis 1:14–16 explicitly stated the celestial bodies to be signs, these texts could not be denying celestial signification. To be consistent, Tycho's opponents would also have to interpret King Asa's being rebuked for trusting in physicians as a condemnation of medicine.[22] No one denied that medicine was a gift of God; no one believed the Scriptures condemned its use. Why not read this text as a condemnation of medicine if the texts cited above are interpreted as prohibitions of astrology? Tycho himself explained that King Asa was condemned for "an excessive confidence in created things if the Creator is also neglected"—this was not a wholesale condemnation of medicine. Tycho simply asked his audience to make a similar distinction with regard to prophetic texts touching on astrology. Celestial influence is not denied nor astrological practice prohibited, only abuse that excludes the power of God.[23]

Tycho had not only to deal with exegetical questions but to correct apparent empirical problems with astrology since he claimed that astrology enjoyed the same status of certitude as other mathematical disciplines.[24] No doubt twins were born under the same celestial influence, but they often experienced very different fates in life and revealed quite different personality traits. How could astrology face these obvious contradictions of its claim that the stars determined such features of a person's life? A prime counterexample could be found in the Bible itself in the lives of Jacob and Esau. The original narrative in Genesis and later biblical commentary on it all stress the opposite traits and fates of these two figures.[25] Tycho's answer invoked a distinction between astral influence and a determinism attributed to astrology by its opponents. While Tycho maintained that "the whole heavenly constitution is in both [twins]", he also insisted that it "varies in diverse ways."[26] Diversity of personality or circumstances of life, therefore, do not suggest that astral influence does not exist but that this influence is worked out in conjunction with the free choices and dispositions of individuals. As for Jacob and Esau, there is more than meets the eye. Tycho quoted the prophet Malachi's statement, "Jacob I loved, Esau I hated."[27] No astrologer would ever claim that the stars bound God to love or not to love someone. "The secret counsel of God" is at work here so that the fates of the patriarchal twins results from "the special will of God alone."[28]

According to Tycho, then, astrology was sorely maligned because it was so misunderstood. It did not imply a fatalistic determinism, as its opponents thought, nor was the true practice of the art to be confused with what the common fortune-telling riffraff did. True astrology assumed God's intention to employ secondary means to influence and shape the lives of his highest creation, man.[29] If, as all Lutherans believed, God was the providential ruler of the heavens, he surely could have used those heavens for ultimate humanitarian purposes. Further, Scripture confirmed this when Moses taught that the heavenly bodies were made as signs for human beings. Everyone acknowledged that celestial bodies were essential for keeping track of the "times and seasons" as Genesis 1:14 asserted. It was precisely that text, Tycho argued, which also justified astral influence.

Why did Tycho take the Genesis reference to luminaries as signs to legitimate astrological inferences and attempt to explain the prophetic prohibitions as misunderstood? Opponents of astrology took the prophetic passages as primary and no doubt would have explained Genesis 1:14–18 as teaching only that celestial bodies mark time for us. Here a classic problem in biblical interpretation shows itself again. Hermeneutical disputants take different passages in the Bible as controlling the interpretation of the issue at hand. No one admits that the passages are contradictory. Each text must be fit with the others, but the question of which text to take as foundational is not easily answered, nor even how to determine the proper interpretation of the foundational passage. Tycho interpreted Genesis 1:14–18 maximally, with its language of signs being a justification for a doctrine of celestial influence; he interpreted the prophetic texts minimally, with their condemnations as not applying to the same doctrine. However, his opponents took the Genesis text minimally, implying only temporal inferences as legitimate, while they interpreted the prophetic texts maximally, so that condemnations of any celestial influence were inferred.

Which method of interpretation was correct? Tycho's lecture certainly did not answer that question, nor does it even seem possible that he could have given an answer at all. These differences were repeated over and over again in the interpretative disputes of the Reformation era without resolution. Niels Hemmingsen apparently did not press the issue because Tycho recorded only that his primary concern centered on the issue of God's freedom, not on the specific interpretation of texts. Tycho's answer to the theodicy problem seemingly reassured Hemmingsen.

Tycho's rehearsal of ancient astronomy among the patriarchs reveals much about his conception of whether the Scriptures contain information to be taken into account by the astronomer. In his later correspondence with Christoph Rothmann (ca. 1589) he reasserted that the knowledge of astronomy was probably passed on to the writing prophets "from the first patriarchs."[30] This allowed him to claim that accurate information about the heavens can be found in the Bible, although the writers were concerned about more important mysteries than astronomy or physics per se. A pattern emerges in Tycho's early biblical interpretation that will characterize his later hermeneutics as well. He thought the Bible was relevant to any subject, but he was also eager to argue that proper interpretation would not imply censure of any science. The beautiful harmony of the heavens and its motions sufficed to declare the glory of God, but recognizing celestial influences, particularly, showed divine care for humans living here in this inferior part of the universe. As Tycho moved toward his geoheliocentric system, he did so with the Bible in the back of his mind, ready to probe its meaning for the sake of science.

Already in 1574, well before his observational program began in earnest, Tycho's interpretation reflected a multifaceted and nuanced approach to the problem of reading the heavens and the Bible. He clearly saw the Bible as a source of history about ancient astronomy— a view not unlike that of many of his contemporaries, who viewed the Bible as a source for virtually every discipline. While never viewing the Bible as an astronomical text, Tycho wanted to ground astrological inferences in the authority of the sacred text. His method for accomplishing this goal invoked the notion that the Bible itself was a source for astronomical information and by implication for astrology as well. Since he believed that astronomy was essential to and found its fulfillment in astrology, he had to overcome theological objections to the latter by showing the Bible relevant to the former. Was this view of the Bible important for Tycho's total restoration project? To answer this, it is important to understand how Brahe arrived at the system that bears his name.

THE DEVELOPMENT OF THE TYCHONIC SYSTEM

The last two decades of the sixteenth century saw several independent attempts to develop a world system that used Copernican parameters within a geostatic framework, but it was only Tycho's system that continued to enjoy prominent acceptance among other learned scholars of Europe.[31] A new planetary system was not Tycho's only goal for astronomy but it was central to the task, for no system could hope to be adequate that did not properly predict the positions and motions of the planets. How did Tycho come to this system?

The first publication of Tycho's system came in 1588 in the eighth chapter of his *De mundi aetherei recentioribus phaenomenis,* a treatise on the comets that Tycho and his assistants had systematically observed. The world system contained in it was only a sketch, and Tycho promised a fuller account in the future.[32] In *De mundi* Tycho claimed to have worked out his system "four years ago"—1583, if 1587 is the year of composition—but there are no indications of his path to the system.

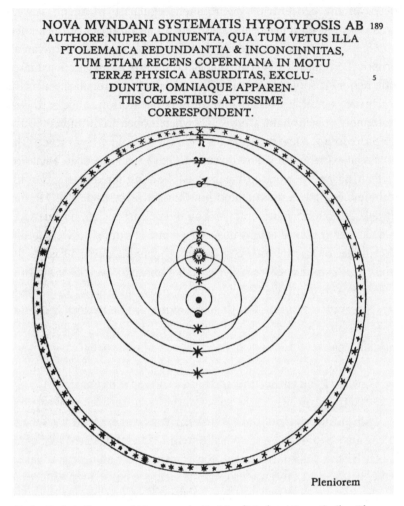

NOVA MVNDANI SYSTEMATIS HYPOTYPOSIS AB 189
AUTHORE NUPER ADINUENTA, QUA TUM VETUS ILLA
PTOLEMAICA REDUNDANTIA & INCONCINNITAS,
TUM ETIAM RECENS COPERNIANA IN MOTU
TERRÆ PHYSICA ABSURDITAS, EXCLU-   5
DUNTUR, OMNIAQUE APPAREN-
TIIS CŒLESTIBUS APTISSIME
CORRESPONDENT.

Pleniorem

Tycho Brahe's diagram of his system in *De Mundi Aetherei Recentioribus Phaenomenis* (1588). Courtesy of the Notre Dame Library.

Tycho divulged something of the path in a letter of the same year to
Caspar Peucer (13 September 1588). At only seventeen (1563), Tycho's
nocturnal activities convinced him of the observational inadequacies
of both the Ptolemaic and Copernican systems.[33] Deeper study of the
Ptolemaic system raised several problems, all of which depended on a
sense of systemic elegance. Tycho was searching for a more natural explanation as to why the orbits of the superior planets (Mars, Jupiter,
Saturn) and the inferior planets (Mercury, Venus) had the sun as their

focus. In the geocentric system, such regularity made no sense.[34] Tycho's search for a necessary cause and a natural order of the planets (*combinatio naturalis*) made him look seriously at the Copernican alternative. Nonetheless, according to Tycho, the modern choice fared little better. It solved some of the mathematical irregularities of the Ptolemaic system, but its attribution of triple motion to the earth was thoroughly objectionable, "especially when it openly contradicts Sacred Scripture in not a few places."[35]

As he explained his development in the letter to Peucer, Tycho referred to a vague period of time in which he considered both sets of hypotheses, but we have no idea how long this period was. He did, however, place enormous emphasis on the year 1582 when he attempted to make an empirical test by observing Mars at opposition to the sun. At the time of this letter (1588) Tycho reported a greater parallax for Mars. This experiment, together with his observations of Venus up to February 1587, compelled him "to give greater and greater credence to the Copernican system," but the continuing problems with terrestrial motion, both physical and theological, forced Tycho to consider yet a third alternative:

> So when I had turned this matter over in my mind in various ways, I carefully considered the [possibility] of another set of hypotheses [*de alia hypothesium ordinatione*]. At first, I seemed to struggle against impossible odds when unexpectedly it occurred to me that if the sun is situated as the center of the five planets and yet revolves yearly around the earth which is at rest with the orbits moving around [*per*] the center and circumference of the five other planets so that the earth has preference over the revolutions of the sun, moon and all the other including the eighth sphere, and if, I say, the celestial revolutions are ordered in this way, then everything that occurs in the Ptolemaic and Copernican assumptions as incongruous, useless and superfluous is removed and avoided.[36]

Tycho proceeded to explain how the notion of solid spheres presented an obstacle to him and how he eventually overcame this problem by incorporating his work on comets. His planetary system demanded the intersection of the Martian and solar orbits, a unique innovation that would have been absurd to most in his time. When he also concluded that the comets must be supralunar, he and many other astronomers had even greater reason to abandon the solid spheres. This of course opened up the problem of celestial matter, a problem to which he devoted much effort in correspondence.[37] For now, it is sufficient to note Tycho's confidence in the truth of his system.[38]

The story of Tycho's development is more confused than his letter to Peucer suggests. His rejection of Ptolemaic inelegance reads consistently from the 1574 *Oratio* to the 1588 *De Mundi*, but his attitude toward the Copernican system is not so clear. He embraced its mathematical superiority, but the physical reality of a moving earth was consistently problematic in Tycho's judgment. In general, he knew that both systems had roughly the same predictive accuracy. However, since Tycho believed that there should be some observational difference between the Ptolemaic and the Copernican systems, he searched for an empirical test, a crucial experiment that would decide between them. That is surely why the 1582 Mars opposition seemed ideal to him, and it is also where the interpretative problem begins for the historian. It was ideal because, in Ptolemy's model, Mars is always farther from the earth than the sun is while in Copernicus's model, Mars at opposition is closer to the earth than the sun is (see diagram). If Tycho could discover which celestial body was closer to the earth at opposition, he could then decide which planetary system was the true one. The only way available to Tycho to measure the Martian and solar distances was by measuring the diurnal parallax. The celestial body with the larger parallax would be the one that was closer to the earth (the larger the parallax, the shorter the distance). Ptolemy's model predicted a smaller parallax for Mars than that of the sun while the Copernican system predicted a larger parallax for Mars at opposition.

Tycho with his assistants made measurements of Mars's parallax on 26 and 27 December 1582 and on 17 January 1583. Like Copernicus, he accepted the ancient value of 3' for the solar parallax so that he expected Mars to display a parallax of about 5' if the Copernican system were true. After a careful reduction of his observational data, Tycho arrived at a negative parallax of 1', or in effect zero. Consistent with these results is Tycho's own claims in a 1584 letter to Henry Brucaeus in which he states that the parallax of Mars was far less than the sun's and that there-

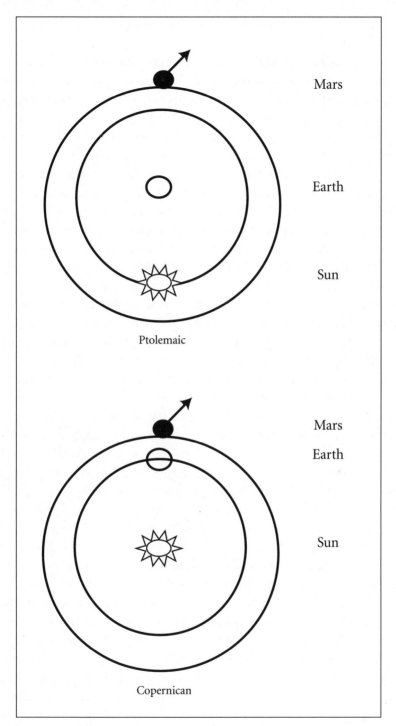

Diagram illustrating Martian and solar orbits in two systems.

fore Mars must be farther from the earth than the sun. These measurements had two effects on Tycho. Since he believed that there should be an observational difference between the two systems, he now also believed, as a result of his observations, that the Copernican system had made the wrong prediction.[39] This did not mean, however, that Tycho embraced the Ptolemaic system. While he may have suffered some bewilderment about the Copernican system, he did not give up on it altogether. What this confused situation seems to have done was to spur him on in his search for an adequate cosmological alternative.

The conundrum for the historian arises because of the contradictory claims about the Martian parallax in his 1584 letter to Brucaeus and in his 1588 letters to Peucer, to Wilhelm IV, and to Rothmann. To Brucaeus he claimed that Mars lay farther from the earth than the sun did, contrary to Copernicus. In his later letters, he claimed that Mars lay closer to the earth in accord with the Copernican model. Many explanations have been offered, and until recently none has proven satisfactory.[40] After four centuries, Gingerich and Voelkel have offered an explanation which surpasses all previous attempts in detail and coherence.[41] They have shown, through a meticulous survey of Tycho's logbooks, that the search for measuring Martian parallax was a sustained and driving motive for the work of the Uraniborg observatory from 1582 to 1587.

The negative results of the 1582 opposition did not deter Tycho. When the next opposition presented itself in January 1585, he was ready to attempt a second data reduction. As Gingerich and Voelkel have shown, these results were even more discouraging than the 1582 results because they suggested a "nonsensical negative parallax." Rather than causing Tycho to abandon his program, these results seem to have compelled him to consider the possible effects of refraction.[42] As a consequence, he constructed new instruments and rebuilt old ones so that he could detect the influence of refraction on his observations. By March 1587 Tycho was ready to measure the Martian parallax again, this time with a large equatorial armillary which he had set up in the new Stjerneborg observatory nearby. Between 1585 and 1587 Tycho made several independent observations to ascertain the effects of refraction on light coming from the stars which he then applied to his Martian problem. Although these measurements were ultimately flawed, Tycho's values for solar refraction became the basis for his comparison with his Martian measurements.[43] When Tycho and his assistants made the necessary corrections for the effects of refraction on the 1587 Martian observations, he came out with a diurnal parallax of 5 and ¾', a number that accorded with the Copernican prediction. Now Tycho had his desired results which he could honestly use to claim that the Martian parallax was greater than the sun's, and that therefore Mars was closer to the earth. His 1587 results explain how he could claim the superiority of the Copernican model in 1588 and 1589. How could he claim that the 1582 observations showed these same results when they clearly did not? Was Tycho intentionally prevaricating when he indicated in his 1588/89 letters to Peucer, Wilhelm IV, and Rothmann that he had obtained these results in 1582? After the 1587 adjustments for refraction, Tycho knew that the 1582 measurements would have shown the same results if he had possessed the increased understanding of refraction made possible by the arsenal of instruments available to him in 1587. He no doubt considered himself justified in claiming 1582 as the date for the greater Martian parallax even though he had failed to find it at that time. In the normal course of research, it would have been sufficient for him to point to the superior observations of 1587, but these were not normal times for Tycho. By this time, he was embroiled in a priority dispute with Nicholas Reimers Baer (Ursus). In October 1588, Ursus published a cosmological system quite similar to the one Tycho had published six months before. Since Ursus had visited Hven in 1584, the unaware reader might conclude that Tycho had learned or, worse, taken, his model from Ursus. In Gingerich's and Voelkel's interpretation, Tycho had the need to establish the priority of his system in the face of Ursus's claims to originality, so he needed to claim his positive results for 1582 to show that his system was not dependent on Ursus in any way. Whether this conjecture is true, Gingerich and Voelkel have amply demonstrated how the search for Martian parallax constituted a centerpiece of the observational program of Uraniborg.

The evidence that Gingerich and Voelkel have uncovered must be read within a wider range of textual evidence than what pertained strictly to the Mars problem, if we are to understand Tycho's development to his system of 1588 in a cosmological context. His earliest statements about Copernicus in the 1574 *Oratio* and his various epistolary indications suggest that the Copernican system was a central concern for Tycho from the start of his astronomical career. Tycho had conceived of his project for the restoration of astronomy by 1574 when he delivered his lecture at Copenhagen and had some vague notion of converting Copernican parameters into a geostatic model, a task he would have also adopted from his German education and contacts. In the lecture discussion, Tycho commented on his own intention to "refer the Copernican spirit and numbers to the stability of the earth," and there are letters of 1576 from Pratensis and Danzeus that mention Tycho's intended conversion project.[44] Tycho was just as sure that Copernicus's attribution of triple motion to the earth was physically impossible and wrongheaded, a belief that remained throughout his life. The scriptural references to a stable earth only confirmed this physical conclusion. During the same period, the new star of 1572 and the comet of 1577 certainly caused him to abandon the Aristotelian doctrine of celestial perfection with its consequent denial of any generation in the ethereal realm. He maintained his belief in a supra/sublunar distinction, however, but not for Aristotelian reasons. Rather, his argument for celestial influences in the *Oratio* suggests that he viewed the cosmos in a Paracelsian correspondence theory that demanded such a distinction.[45] Retaining this belief also fit well with his continuing belief in solid spheres. At the close of the 1570s, Tycho still had only conceived his project in very broad outlines, but even at this early stage his program had a Copernican flavor to it, so much so that it is possible to call it, as Gingerich and Voelkel have, a "Copernican campaign."

Paul Wittich arrived on Hven in the summer of 1580 when, as Gingerich and Westman have conjectured, he outlined his own model for transforming Copernican parameters into a geostatic model.[46] This would have been very attractive to Tycho because he could see no physical or theological reasons to posit a moving earth. He repeatedly told the recipients of his letters that a moving earth was against physics and Holy Scripture. Wittich's model would have also made Tycho realize, however, that the sun could be the center of the five planetary orbits, both superior and inferior. This realization would also have reaffirmed his basic Copernican instincts and made him search even more diligently for a way to confirm the Martian parallax because he saw that his own system, like Copernicus's, required an approach of Mars closer to the earth than the sun's approach at opposition. Tycho's tenacity in finding the Copernican numbers for Martian parallax is understandable if he viewed his own system as standing or falling with this feature of the Copernican system.

Another problem no doubt presented itself sometime in this process. In Wittich's model, Mars's orbit completely encircled the sun. Tycho must have seen that this model violated the ratio of the sun's orbit to that of Mars, namely, 2:3, and that Wittich's model needed to be adjusted to allow for an intersection of the Martian and solar orbits. Up to this point, all of Tycho's ideas could have been interpreted within a system containing solid spheres, but once he realized that the true system would require an intersection of spheres, he was probably quite perplexed. The comet of 1585 must have been a welcome sight because it forced Tycho and many other astronomers to rethink the existence of the solid spheres. Yet it had a special significance for him that it did not hold for others like Maestlin, Ursus, or Rothmann. He realized that he could now admit the intersection of the Martian and solar orbits because there were no solid spheres. As he said to Peucer later, his confidence in his own system grew dramatically.

This scenario suggests that the emergence of Tycho's system was extended, tedious, and piecemeal. Gingerich and Voelkel have shown, at least to my satisfaction, the pervasive Copernican motivation that sustained Tycho's program. I have suggested that the early evidence from the 1570s accords well with their focus on the Mars problem of the 1580s. Of course, he could never have seriously proffered his world system as a true cosmology unless he had solved the problem of the intersecting spheres. The cometary observations of 1585 and the Martian observations of 1587 gave Tycho what he needed to claim that his system solved the astronomical problems of Ptolemy's system as well as the physical and theological problems of Copernicus's. To his mind, Tycho had worked out a system that met all the demands of observation, natural philosophy, and scriptural interpretation. In agreement with physics and Holy Scripture, his system had the earth at rest in the center of the universe. The sun was the center of five known planets, a feature that enhanced the systematic aesthetics implicit in Tycho's astronomy. Most importantly, however, his system was observationally superior to both major contenders because it allowed for Mars's orbit

to intersect the sun's orbit; this was possible only because he had defini-
tively disproved the existence of solid spheres. Tycho had great confidence in his system, as his dispute with Ursus showed, and he believed
it to be not only better than other geoheliocentric systems of the late
sixteenth century but the true system that God had created. In his own
mind, he had truly combined astronomy, physics, and Holy Scripture
into perfect agreement.

TYCHO, THE BIBLE, AND THE MOTION OF THE EARTH

Explaining Tycho's belief that Scripture taught an immobile earth
proves somewhat difficult because his citation of scriptural opposition
to earth's motion is never accompanied by specific texts. He often
spoke of the earth as a "sluggish and ignoble body" that was unfit for
the velocity needed to account for diurnal motion. Whenever Tycho
gave arguments against the motion of the earth, he always claimed that
it was against physics and Sacred Scripture:

> For although it [Copernican theory] conveniently remedies those
> other things which in the Ptolemaic system are incoherent and su-
> perfluous and it lacks nothing that is mathematically good, never-
> theless, when it attributes a regular, perfect and less intricate
> motion to the earth (that sluggish and ignoble body), this assump-
> tion is rendered no less suspect, especially when it openly contra-
> dicts Sacred Scriptures in not a few places.[47]

The tone of Tycho's objections to Copernicus makes clear his certainty
about what can be known from physical considerations and from the
biblical teaching. With such a strong conviction, it would appear natural for Tycho to debate the meaning of texts that were subject to alternative interpretations. But he did not do this. He accepted the biblical
statements about a motionless earth without deliberation or dispute,
and he seemed unaware that the texts dealing with terrestrial motion
were open to alternative interpretations. In fact, he never quoted or
even cited any standard texts (Ps. 93; Eccles. 1; Josh. 10)—a perplexing
lack of comment because some of his correspondents (e.g., Rothmann)

had raised the very issues that demanded such discussion. Tycho's lack of comment is all the more puzzling when we consider that he was ready in other instances to dispute interpretations of Scripture he thought wrong. We observed above how Tycho departed from interpretations of the prophetic texts that were taken as proofs of the illegitimacy of astrology.

Tycho's lack of hermeneutical discussion can be explained by two factors: his view of disciplinary boundaries and the structure of his cosmology. His differences with Christoph Rothmann over the proper relations between astronomy, physics, and theology were both profound and highly consequential. Rothmann argued that physics or theology could not answer the question of celestial matter, but the form of his argument suggests that he would certainly have said the same with regard to the Bible's relevance to the earth's motion:

> Unless this question [celestial matter] is decided by us, it will not be decided by anyone, whether theologian or physicist. For God has not revealed anything whatever about this in his Word because it has nothing to do with our salvation. The Scriptures, which are written for the unlearned and learned alike, the common and ingenious, do not contain such disputations which are not even understood by very many learned, as Christ testifies in John chapter three to Nicodemus.[48]

For Rothmann, the Bible is a book of redemption, not natural philosophy, and he appealed to the notion invoked by Rheticus that was widely known among scriptural interpreters, accommodation. Yet Rothmann's categorical exclusion of biblical relevance to natural philosophy appears more thoroughgoing than that of some earlier interpreters. His reference to the story of Jesus and Nicodemus might imply that for Rothmann the Bible is not a book of disputations at all, not even theological disputations. Rothmann probably viewed Nicodemus as wanting to engage Jesus in rabbinic-style disputation while viewing Jesus as rejecting this style in favor of deeper spiritual truths. From this, he inferred Jesus' intention to be edificatory, not theological. Generalizing this intention to the entire Bible—an obvious nonsequitur—would have had persuasive force only against the background of a common belief in accommodated scriptural language, a belief Tycho shared. Rothmann probably reasoned that if the Bible was written for spiritual edification and not for theological disputation, then certainly it was not written for natural philosophy. That the Bible was written for everyone in everyday language precluded its being interpreted as a textbook for arcane philosophy.

If theology had no relevance to the question of earth's motion, surely physics did, since kinematics constituted a well-recognized part of this science. However, physics cannot yield an answer for quite different reasons:

> Also, how can the physicists know anything with certainty? For we know and understand about the heights and the matters discussed by us only as much as we discover mathematical demonstrations through trigonometry. Without these [demonstrations] those who discuss such matters are completely worthless and raving mad.[49]

Rothmann analyzed Aristotelian physical claims into one of two categories—either they constituted metaphysical claims that could be "as easily denied as affirmed" or they should be construed as empirical claims requiring verification.[50] The former category had no capacity for proof, and the latter lacked sufficient demonstrations. The only possible answer for Rothmann lay with astronomy, because it provided answers that were both relevant and demonstrative. The priority of astronomy in establishing knowledge of the heavens must ultimately be behind Rothmann's sudden adoption of Copernicanism.

Tycho's view of the relevance of physics and theology was quite different. His underlying assumption was that discovering the true system of the world required the agreement of all three disciplines in their respective conclusions, but this did not imply that all disciplines were relevant to the same degree:

> Therefore the question of celestial matter is not properly a decision of astronomers. The astronomers labor to investigate from accurate observations not what heaven is and from what its splendid bodies exist, but rather especially how all these bodies move. The question of celestial matter is left to the theologians and physicists among whom now there is still not a satisfactory explanation.[51]

What Tycho said about celestial matter also applies to terrestrial motion. In the hierarchy of the sciences, the question of motion was a physical question, although astronomy still had a limited role to play. Astronomy could construct tests to refute the motion of the earth, but no amount of observation or theoretical construction could positively establish the earth's motion because motion could only be attributed to bodies whose nature we know. Astronomy did not deal with the *natures* of bodies, celestial or elementary; it only observed and hypothesized about celestial motions. Physics, on the other hand, could provide a positive answer, and in Tycho's view no physical reason for departing from the standard view of terrestrial immobility existed. His references to earth's immobility always retained the order of being contrary to "physics and Sacred Scripture"—he never reverses this order. This seemingly insignificant detail becomes important and explicable when his views on the relations of the relevant disciplines are taken into account.

I suggest that Tycho's belief in an implicit hierarchy explains his lack of comment on biblical passages dealing with the motion of the earth. If terrestrial motion could not be settled by physical arguments, then Tycho would have had to consider theological arguments and handle alternative interpretations of texts. If, on the other hand, the proposition of terrestrial motion was denied by physics, then the question was settled and there was no need to resort to an analysis of biblical data.

That Tycho believed the issue was settled within the realm of physics finds support in a letter to Kepler toward the end of Tycho's life (1598). This missive offered a critique of Kepler's Copernicanism while praising the ingenuity of the *Mysterium Cosmographicum*. As he had from the beginning of his investigations, Tycho reiterated his standard objections to the Copernican system, including yet another denial of the triple motion of the earth:

Nor could the earth have an annual motion so I can completely ignore superfluous librations (nor is there any such thing as a precession of the equinoxes as Copernicus thought). However, a diurnal motion can be plausibly attributed to it because of a smaller rotation [*gyrum*] but while this would fit so heavy and dense a body [earth], I adjust [this body] to rest rather than to motion.[52]

Missing from Tycho's comments is his usual biblical denial of terrestrial immobility. Was this an oversight or was it no longer relevant? In their earlier forms, Tycho's denials rarely mentioned the three kinds of Copernican motions specifically; they simply denied any and all movement. To Kepler, Tycho completely denied annual motion and libration while showing some allowance for daily motion. Ten years earlier, in making his system public, Tycho strongly denied any motions to the earth, although some other geoheliocentric systems had incorporated diurnal motion. At that time, he needed to argue against any terrestrial motion because his own system was not yet widely accepted. Perhaps by 1598 Tycho perceived that his system (or some geoheliocentric variant) was gaining acceptance, and since some other semi-Tychonic systems had allowed for diurnal motion, he might have left the question open. He would then have had to reconsider his biblical opposition, and that would explain his lack of reference to terrestrial motion to Kepler. On the other hand, it is also clear that Tycho still thought of the earth as immobile and he may not have cited biblical reasons any longer because his confidence in the reality of his own system had still convinced him of the physical impossibility of the earth's motion. Now that he possessed his own alternative to Ptolemy and Copernicus, it would not be as necessary to emphasize confirmatory evidence from the Bible. How ironic that Tycho's confidence would later be undermined by the recipient of his letter, the very one to offer physical reasons for the earth's movement.

The structure of Tycho's cosmology also affords insight into his view that the earth cannot be moving. In general, Tycho assumed that the elementary world consisted of the four elements (air, earth, fire, water), and he believed that nothing in the elementary, sublunar world could be ascribed to the celestial realm.[53] Rothmann argued that air permeated both the sublunar realm and the heavens all the way up to the eighth sphere (of fixed stars).[54] The reasons for Tycho's disagreement were already evident in his *Oratio* of 1574. His argument for celestial influences, noted earlier, depended on a rigid separation of the celestial and elementary worlds. The sublunar realm, made up of only air, received the influences of the heavens and transmitted these influences to human beings.[55] Tycho drew upon the Paracelsian correspondence between planets and human organs to argue that human beings consisted of the four elements and also participated in the forms of the superior world.[56] Such notions of the universe and humans assumed, and sometimes asserted, a distinction between the two realms that demanded an argument for a connection between them—a rigid distinction common both to Aristotelian and Paracelsian cosmologies.

Consequently, there was no need in Tycho's mind to argue between variant interpretations of biblical texts on terrestrial motion because there were no compelling physical reasons to believe that anything of an elementary nature participated in the celestial region. To reverse the positions of the sun and earth, as the Copernican system did, would violate this cosmic structure that God had placed in the world. The earth could not participate in celestial motion because of its nature. Knowledge from physics and the evident meaning of the Bible agreed— the earth could not be moving. But Tycho's discussions of the Bible on other matters suggests that he would have seriously reconsidered those biblical texts on terrestrial motion that seemed so clear if he had physical reasons to do so. His willingness to depart from literal physical interpretations of biblical texts becomes evident in his discussions of celestial matter.

TYCHO, PEUCER, AND ROTHMANN ON THE BIBLE AND THE CELESTIAL REALM

The greater part of Tycho's biblical exegesis concerned the problem of the nature of the celestial realm in general. Once the doctrine of the celestial spheres had been abandoned by many astronomers (ca. 1580s), they found themselves obliged to address the question of what made up the heavens. The wide range of answers posed even deeper problems of how to adjudicate among these solutions, and it is here that the interrelations among disciplines had a direct effect on the theories proffered. Among the variety of proposals for celestial matter, many had roots in ancient philosophy. Naturally enough, one could find traditional Aristotelians who maintained a doctrine of the inalterability of the heavens by explaining the comets as meteorological or visual phenomena although even the most devoted disciples of the Stagirite displayed a greater flexibility than is generally acknowledged.[57] One prominent trend was the Stoic revival of a continuous view of matter encompassing both the celestial and terrestrial worlds. Under the influence of Jean Peña, the mathematician of Paris, this view took the form of the heavens consisting of air.[58] Nor were the chemical philosophers silent on the matter. Paracelsus had spoken of the heavens as having an igneous substance.

Tycho was well aware of these various developments. As noted above, he felt that he could not turn to the physicists or theologians for a satisfactory answer to this question, so he was compelled to enter into each arena to determine as best he could the proper answer. In this process, he contemplated various theories—some ancient, some modern. Because he lacked a definitive answer from natural philosophy, scriptural declarations on the heavens became relevant; he had to evaluate not only natural-philosophical positions but weigh various interpretations of texts. What was his methodology in this process? How did this affect his interpretation of texts which seem to teach the solidity of the celestial spheres? How did he interpret texts implying a doctrine of celestial matter? I shall take up each of these questions in order.

I suggested earlier that Tycho's literalism was only apparent and that his hermeneutical complacency with regard to terrestrial motion grew out of his prior belief in its physical impossibility, not from a naive biblicism. The implications for Tycho's cosmology surfaced when he interacted with cosmological alternatives of his day, one of which was suggested by the Spanish physician Francisco Vallés in his *De sacra philosophia* printed in Leiden in 1588, the same year in which Brahe sent his lengthy letter to Peucer detailing his path to his system. We shall return to Tycho's critique of Vallés momentarily, but first we note that in Tycho's judgment the faux pas Vallés commits involves a too literal interpretation of Scripture:

> But he [Vallés] understands this text and certain others in a much too literal fashion. Nor does he consider what is said that is contrary to this in Isaiah and elsewhere.[59]

> [Vallés' method] propounds things too much from the literal sense of Scripture. Let us abandon this way if indeed it leads us away from our present duty and if it is resolved to investigate with those

FRANCISCI
VALLESII,
DE
SACRA PHILOSOPHIA,

SIVE DE IIS, QVÆ IN LIBRIS
facris phyſicè ſcripta ſunt,

LIBER SINGVLARIS,
Cui ſubiunguntur duo alij , nempe
LEVINI LEMNII *de Plantis ſacris,*
E·T
FRANCISCI RVEI *de Gemmis.*

EDITIO SEXTA
Correctior, & iuxta Indicem expurgatorium reformata.

LVGDVNI,
Sumptib. IOANNIS ANTONII HVGVE
& MARC. ANT. RAVAVD.

M. DC. LII.

Title page of Francisco Valles's *De Sacra Philosophia*, 1652 edition. Courtesy of the Notre Dame Library.

who do chiefly astronomy the many things contained in this science. So there can be no objection from Sacred Scripture or philosophy if we established with certainty that the material of the heavens is very liquid and more tenuous and subtle than any air.[60]

Tycho's criticism sounds akin to what Galileo and Kepler would later argue, namely, that the issues of physical science and biblical interpretation are separate and that one should not rely on a literal reading of the Bible to arrive at a picture of the universe. Yet Tycho's language cited earlier with regard to earth's motion ("against physics and Holy Scripture") suggests that for him the Bible was still relevant to physical questions. This ambivalence on Tycho's part suggests the need to analyze his hermeneutics with a finer stroke than has previously been attempted.

To Christoph Rothmann, Tycho must have seemed like a hidebound literalist, for Rothmann objected to dragging Scripture into an essentially astronomical discussion. He claimed that the biblical language was gauged (accommodated) to the understanding of common people who did not have access to esoteric astronomical knowledge. It cannot therefore be expected that the Bible will yield information helpful to the astronomer. Tycho's response admitted that the Bible uses a common method of description in scientific matters but that this does not imply that its words must not be taken seriously. For Tycho, this would be treating the Bible as if it were another human document that need not be taken authoritatively when speaking about astronomical matters:

> Much less do the things you affirmed deserve a place because you excuse those things that Holy Scripture asserts to the contrary. The reverence and authority due to the sacred writings is and ought to be greater than that of dragging them into common discussion. For although they adjusted themselves to the common method of understanding in physics and some other matters, yet let it be far from us to think of them as speaking in *such* a common manner that we do not believe them to be speaking truth. Thus Moses, even if he does not refer to the deep things of astronomy when treating the creation of the world in the first chapter of Genesis, because he

is writing for the common people, nevertheless he does introduce that which our astronomers can concede.[61]

In Tycho's estimation, Rothmann's dismissal of biblical relevance effectively leveled the authority of the Bible and put it on the same plane as other physical treatises. Tycho insisted on the potential relevance of the Bible to any physical question, since to deny it would preclude the possibility of divine authority answering the question. Similarly, he praised Caspar Peucer for confirming from Scripture that the heavens are not made of solid spheres.[62]

The use of phenomenal language in the Bible did not imply that its authority was limited, only that its information was not complete. Tycho's discussion of Genesis 1:16 makes this clear. This text spoke of two great lights in the heavens, the greater (sun) and the lesser (moon). Rothmann had interpreted this, in keeping with much historical precedent, as accommodated language since geometrical demonstrations proved that the moon was not greater than the other stars. Rothmann did not deny the truth of the Bible but only that its language could be taken literally (thus also with regard to the earth's motion). Tycho's words above show clearly his recognition of accommodated language, but he also insisted that one could interpret the language of Genesis 1:16 in a physical sense if one takes it as referring, not to the size of the body, but to the magnitude of light. To defend this position, Tycho offered a counterfactual argument:

> if he [Moses] had meant this about the magnitude of the bodies themselves and especially had accommodated himself to the common view, he would not have called the moon less than the sun since the visible diameter of the moon is hardly less than the sun; sometimes it is greater.[63]

Although the language of Genesis 1:16 is gauged to an earthly observer, it is not contrary to physical truth since it refers to the appearance of light, not to the body itself. For Tycho, Rothmann had needlessly impugned the authority of Scripture simply because he did not understand its language properly. Tycho's objections to Vallés and Rothmann point to a subtler hermeneutical methodology than simple physical literalism. Against Vallés, Tycho denied any authoritative natural philosophy in the Bible (a "sacred philosophy") but, against Rothmann, he insisted on the relevance of the Bible because in the prophetic writings "many things drawn from physics are mixed into their oracles."[64] This hermeneutic stands between an attempt to settle questions of natural philosophy by simple appeals to the literal meaning of the Bible and a dismissal of biblical relevance based on an argument from accommodation. Tycho attempted to cull from the Bible any information that might have bearing on the problems he faced, but he did so without any illusion that he could found a total cosmology on biblical texts.

This approach affords insight into Tycho's handling of texts that bear on the issue of solid celestial spheres. The new star that appeared in the constellation Cassiopeia in 1572 convinced Tycho of the alterability of the heavens, since this new phenomena had never before been seen or even appeared. However, it was the comets of 1577 and 1585 that convinced Tycho that there could be no solid, impervious spheres in the heavens because the comets were clearly above the sphere of the moon (contra Aristotle). Armed with this empirical data, he often reiterated his denial of solid spheres:

> Heaven is not made up of real, durable and impervious orbs to which the stars are affixed and travel, but it consists of a substance that is very clear, very thin and very fine. This makes the courses of the seven planets free so that they move without any slowing wherever their natural impetus and their relations carry them. This was not seen by the ancients or even the greater part of the moderns nor even conceded because it was never doubted. For it is enough for the restoration of astronomy to admit it as settled and known.[65]

Tycho saw this physical truth as a sure sign of the superiority of his system over both Ptolemy and Copernicus because only he possessed the empirical evidence that allowed for the intersection of planetary spheres. Tycho now also knew that the Bible did not teach a doctrine of solid spheres because nothing in the Bible would contradict this irrefutable empirical knowledge. Others were not so convinced and offered biblical evidence for the solid spheres, texts that Tycho had an obligation to interpret. The Greek translators of the Septuagint had translated the Hebrew *raqia* of Genesis, chapter 1, as *stereoma*, which in Greek philosophical usage meant "a solid body."[66] This was translated into Latin as *firmamentum*, which was then taken in the West as evidence for the Bible teaching the solidity of the celestial spheres. Tycho argued that *raqia* is more properly translated by *expansum*, taken from Castalio's translation, and indicates a liquid or an open expanse that allows the free movement of celestial bodies.[67]

His commitment to a liquid heaven required Tycho to address two key texts that seemed contradictory on the surface, Isaiah 40:22 and Job 37:18. In his letter to Caspar Peucer (13 September 1588) Tycho approvingly cited Castalio's version of Job 37:18, no doubt because Castalio had used *aethera* to translate the Hebrew. Castalio's translation was more an insertion of a modern cosmology into an ancient text than an exegesis of its original meaning.[68] A more proper translation of the Massoretic text of Job 37:18 is, "Have you, Job, worked with him [God] to fashion the hard vault of heaven with a hammer so that it looks like a mirror of cast iron?" The impression that this text teaches a solid celestial sphere could only be reinforced by the Vulgate translation of the Hebrew *chazaq* with the superlative form *solidissimi* (= very hard). The text also has the verbal root *rq'* (= to hammer out) which is related to the firmament (*raqi'a*) of Genesis, chapter 1. This verb was regularly used in the Hebrew Bible to refer to the fashioning of metal products. Consequently, a thorough investigation of Job 37:18 only confirmed the initial impression that Holy Scripture taught the solidity of the celestial firmament and thereby justified the translation of *raqia* by the Latin *firmamentum*. Would not this language, if it were taken literally, require belief in a solid sphere? Does Tycho take it so? Neither Tycho nor Peucer viewed this text as demanding a solid sphere, the former commending the latter for his reconciliation of Job 37:18 with Isaiah 40:22:

It [Job 37:18] says that the heavens are hard like steel. This is correctly expounded by you in a learned and proper manner when you reconcile this text with the earlier one by saying that the heavens have a solidity and perpetual firmness of constancy (along with everything in them) rather than referring to the material of which the heavens are made.[69]

Both Tycho and Peucer thought the Bible contained important information on cosmology, and Tycho's interpretation of Job 37:18 did not place it outside the realm of relevance, but neither did it involve a literal description of a physical celestial reality. The words had a meaning that referred to the visible constancy of the heavens so that the solidity mentioned in the text was not accommodated language; it was literal but not physical.

This method explains Tycho's criticisms of Vallés mentioned earlier. Although Tycho complained of Vallés's literalism, the real danger that the Spaniard fell into was his attempt to join Scripture too closely to a "sacred philosophy" based on certain texts. Vallés had woven together such texts as Job 37:18 with quotations from ancient Greek philosophers showing the agreement of Sacred Scripture with philosophical opinion. He concluded that the heavens could not consist of a liquid and penetrable substance but rather must be made up of the four sublunary elements combined with solid orbs. Vallés's literalism went from absurdity to absurdity when he contended that the new star of 1572 was really nothing new but had been there since the creation of the world and that it had simply not been seen due to its small size. Vallés's view was based on Genesis 2:4, where it says "the heavens and the earth and all their hosts were completed and God completed His work on the sixth day." Tycho was incredulous that so learned a man could start from a literal interpretation and arrive at conclusions that lead to absurdities.

> How does he [Vallés] establish short of all sense and reason that this star has been there since the beginning of the world but was too small to be seen? . . . O what speculation! One absurdity is only equaled by another. Heaven has no firm orbs that really exist. Much less are there substances that are rarer in some places and denser in others. (Ibid., 134)

Vallés's literalism must be answered, not by an appeal to accommodation, as Rothmann would have it, but by a closer ascertaining of the meaning of the text. That meaning will agree with irrefutable observations that had long ago been established. The biblical texts then must be subject to a prior assessment based on astronomy, far from a literalism that controlled astronomy. Tycho made the authority of the text dependent on observations of the heavens. This priority no doubt caused him to seek meanings for the texts which were literally true but not necessarily physical.

Denying the existence of solid celestial spheres did not answer the question of celestial matter. Tycho argued against Rothmann's thesis that the celestial region contained air, as the sublunar realm did, but his engagement in interpreting biblical texts came into prominence in his correspondence with Caspar Peucer. His high respect for Peucer's handling of the biblical texts was probably shared by many in Germany and other Lutheran strongholds, but it did not prevent him from deep disagreements over what might be inferred from passages suggesting the nature of celestial matter. Peucer agreed with Tycho that the heavens were made up of "an ethereal substance that is refined, pure and accessible to all rays of light and is liquid and fluid."[70] Both Tycho and Peucer agreed that there should be no confusion of the elementary and the celestial realms, but Peucer misunderstood Tycho's position when he attributed to him the Paracelsian view of the heavens as an igneous substance. Tycho admitted no elementary substance into the heavens.[71] This difference becomes evident in their disagreement over the interpretations of *raqia*.

Peucer held to supercelestial waters "above the firmament" where the eternal light of God dwelt. His grounds lay in the etymology of the Hebrew word for heavens, *schamayim*, which he derived from *sham* = there and *mayim* = water. This etymology was confirmed by the language of the flood narrative speaking of the "floodgates of heaven" (Gen. 8:2) as well as other texts that spoke of "dark waters" and "thick clouds of the sky" (Ps. 18:11). The light that reached the earth came from the eternal seat of God and, when joined with the waters and clouds, produced the sapphire color of the sky.[72] The firm foundation of Scripture was to be preferred to the wranglings and uncertainties of the philosophers.[73] The foundation of astronomy was sure observations, but apparently these could not answer important questions about the nature of celestial matter. For these, one had to turn to the most reliable source of divine revelation, the Bible.

Tycho objected strenuously to Peucer's exegesis. To identify the waters of Genesis 1 as supercelestial would be to claim that Scripture was speaking of things not obvious to every observer:

Moses composed the account of creation for common and simple people who were not acquainted with mathematics and physics and he undertook to explain things which appear to men's eyes, not invisible and obscure things.[74]

On the surface, this appeal to accommodation on Tycho's part is similar to Rothmann's argument against Tycho on the motion of the earth. However, Tycho did not dismiss Peucer's exegesis as irrelevant; instead he offered an alternative reading. He appealed to a greater flexibility of language evident in the Bible. Deuteronomy 4:24 and Hebrews 12:29 spoke of God as a consuming fire, but this had no relation to physical fire or heat that we experience. This was to be seen as metaphorical extension.[75] Similarly, the word *raqia* did not always have the same referent. Sometimes it meant the total expanse of heaven as it appeared to the naked-eye observer. Other times it referred to the supralunar realm of the luminaries and still others to the liquid region just under the moon. Acknowledging this flexibility of linguistic reference in the sacred text allowed one to recognize that Moses was not referring to any putative supracelestial waters: the two waters mentioned in Scripture are referring to the seas on earth *and* the clouds, lightning, and rain which are in the region of the air.[76] In the end, however, Tycho argued against supracelestial waters on theological-aesthetic grounds: "God is the author of order and utility."[77] This implied specifically for Tycho that nothing from the elementary realm, including water, could be ascribed to the supracelestial world where God dwelt.

Tycho's discussion of Vallés was directed to Peucer as a warning not to attempt to base a cosmology on the Bible. Cosmology was founded on observational astronomy and on whatever could be learned from physics and chemical correspondences. The Bible indeed had relevant information but not a complete sacred philosophy. When it spoke of the natural world, it spoke truly and it must therefore be taken seriously. To understand the meaning of the biblical texts that spoke of the natural world, one had to be sure of what could be said with certainty and what was only conjecture. This would prevent one from falling into either Vallesian literalism or Rothmannian irrelevance.

THE NEW UNIVERSE AND THE BIBLE

By the 1590s the universe conceived in the mind of astronomers looked very different from the one Copernicus had imagined, not because they rejected Copernicus but because they no longer believed, by and large, in the existence of the celestial spheres. The only issue in Wittenberg that had any theological significance in the 1540s was the motion of the earth. By the end of the century a whole new set of questions had emerged to which theology seemed plausibly relevant. These questions centered around celestial matter, but they were intensified by the growing sense of realism that was emerging among astronomers. Although Kepler was the first to propose an extensive and explicit physical version of Copernicanism, the path was laid for him by Tycho, Peucer, and even Rothmann as they pursued a cosmology that would unite mathematical astronomy to physics and natural philosophy. In this connection, I would argue that Tycho is truly a transitional figure in that he retained the traditional view of hypotheses as convenient fictions, while at the same time he seems to have believed that one set of hypotheses would in fact be established as the true system. Which set it was to be would depend on the confirmatory conclusions of physics and theology, the only disciplines that, for Tycho, can tell us about the physical nature of the heavens.

Tycho worked on an implicit hierarchy, but not the medieval one in which astronomy yielded to physics and physics to theology. His hierarchy rather located a question in a proper discipline. If it was predicting planetary motions, astronomy was the primary discipline. If it was the motion of the earth or the nature of celestial matter, physics or natural philosophy would make the determination. The other disciplines aided the primary discipline in the task, especially if the question could not be settled in the primary discipline. With regard to natural questions, theology was never the primary discipline, but it could become relevant if physical matters were left unsettled by their own proper discipline. In this hierarchy, any discipline could be primary or ancillary, depending on the question being posed. This Tychonic hierarchy explains Brahe's confidence in his world system: it met all the demands of each discipline. It had observational superiority, and it displayed harmony with the five planets centered on the sun. It agreed with physics by its rigid sub/supra lunar distinction, and there was no objection of Scripture against it.

Tycho's biblical interpretation defies simple categorization, but it is clear that he was no simplistic literalist. That he has been so thought only shows that the subtlety of his interpretations have not been grasped. Historians have sometimes left the impression that those who opposed the mobility of the earth did so because they believed in biblical relevance, a belief that arose naturally from commitment to literal interpretation. Tycho belies this construction. He certainly believed that the Bible had much to say about the universe and that these statements could not be easily dismissed, but he also believed one could not simply go to the Bible and expect to find there a sacred physics. This suggests that historians should not conflate the issue of biblical relevance to cosmology with the issue of literal interpretation. Otherwise, we will miss the richness of strategies employed in resolving apparent conflicts between the new astronomy and biblical interpretation. Tycho's and Peucer's interchange over biblical texts suggests that the Wittenberg tradition had by no means a uniform hermeneutical method. As in most theological traditions, there was interpretative diversity within the parameters of a unified confession.

Kepler, Cosmology, and the Bible

O f all the stars in the early modern pantheon, none shines more brightly than that of Johannes Kepler. His achievements were so numerous and revolutionary that those developments usually summarized under the term "Copernican revolution" can, with considerably more justification, be properly called Keplerian. Although Galileo springs up in the popular mind as the greatest of early modern astronomers, his actual astronomical achievement was meager compared to that of Kepler. It was Kepler who first searched systematically for physical causes of celestial phenomena and whose mathematical application achieved a degree of accuracy previously unknown. It was Kepler, with his formulation of the three laws of planetary motion, who finally abandoned the circle, embraced elliptical orbits, and provided the necessary background for Newton's universal gravitation. If the work of any one seventeenth-century astronomer can be easily extracted from his religious convictions and context, Kepler seems to be a prime candidate.[1] As a premier example of meticulous observation, detailed thought, and stunning mathematical ability, he changed during his life from animistic to physical explanations of the universe. He himself argued that religious considerations should not intrude unwanted into questions of astronomy. More than once he invoked the notion of accommodated language in the Bible to keep at bay those who would undermine the Copernican theory by scriptural appeals.[2] The separation of Kepler's astronomy from his cosmology was already done by early interpreters; the harmonic and Platonic aspects of his cosmology were all but forgotten by subsequent astronomers, even by those who held fast to his physical principles.[3] While recognizing Kepler's deeply religious inclinations, modern interpreters have tended to assign his belief in biblical authority to the moral sphere and his astronomy to the search for scientific truth.[4]

Yet the reader of Kepler's works cannot help but be impressed with the interconnectedness of the mathematical, physical, metaphysical, and theological aspects of his thought. Kepler's desire early in life was to be a theologian, and he later thought of himself as an exegete of God's works in nature. Kepler's words to Maestlin, penned during the writing of his first major work, shows his struggle: "I wanted to become a theologian. For a long time I was restless. Now, however, behold how through my effort God is being celebrated in astronomy."[5] A growing recognition of the centrality of religious ideas and motivations has been emerging in Kepler scholarship over the last two decades.[6] Kozhamthadam's study contends that Kepler's religious ideas played an indispensable role in the discovery of his famous laws of celestial motion. He takes seriously Kepler's numerous expressions of desire to bring glory to God through his astronomy as a motivation for his use of the Trinity-sphere icon as a heuristic concept in the discovery process.[7] Yet there remain many unanswered questions, not the least of which are the complicated ways Kepler uses the Bible and its categories in thinking about nature. Understanding the role of biblical interpretation in Kepler's cosmology requires accurate contextualizing of all these aspects of his thought, regardless of how they were separated by subsequent interpreters. I will argue that Kepler's understanding of his task and his resultant cosmology are deeply informed by biblical categories, while simultaneously rejecting any attempt to make the Bible into a scientific authority.

KEPLER'S SELF-UNDERSTANDING

Kepler's goals, clear and consistent throughout his career, show how deeply Christian theology informed his cosmology. When he published his first exposition of the Copernican theory in *Mysterium Cos-*

mographicum (1596), he was already intent on finding a physical version of the Copernican system that met the requirements of mathematics and physical theory. He conceived of his task as bringing unity and harmony into disparate planetary models because the ultimate source of the world system was the One in whom unity and diversity coexisted, the triune God. The modern reader is now likely to miss the significance of the title, *Mysterium Cosmographicum*. Kepler was later to remark that the direction of his whole life was set by this book because it immediately thrust him into the forefront of leading European astronomers, a direction attested to in the longer version of the title:[8]

> A forerunner of cosmographic writings containing a cosmographic mystery about the admirable proportion and causes of the numbers, sizes and periodic motions of the heavens demonstrated by their own real principles through the five regular geometric solids.[9]

Kepler intimates that this is only the first of several cosmological treatises to follow, all of which would contribute greatly to the reception of the Copernican system.[10] The double occurrence of *cosmographicum* is strikingly unusual for a sixteenth-century work in astronomy. Kepler himself tells us that he labored over the proper title of the book because he did not want it to be understood as rigorous mathematics or as a complete cosmography.[11] He first used the term in a student disputation at Tübingen on the motion of the earth, where he glossed the meaning as "light reasons, a kind of thinly spun cosmographicum."[12] Kepler contrasted these lighter reasons with the kind of mathematical demonstrations normally expected in astronomical discussions and apparently intended to indicate physical considerations. A physical meaning became even clearer in the actual book itself.

But why the term *cosmographicum*, and not *physica* or *sphaerica?*[13] I argued in chapter 1 that the term *cosmographia* derived from the Florentine revival of Ptolemy's *Geography*. Kepler seems to use the word *cosmographia* in Ptolemy's sense of treating the world as a whole, but with one major difference. Ptolemy and the tradition of geography derived from him used "world" to refer to the physical earth, whereas Kepler draws on the ambiguity of *mundus* to refer also to the whole planetary system.[14] His purpose is the same as Ptolemy's, to represent the various parts of that world system in relation to one another. This holistic concern was complemented by a second feature. Ptolemy's *Geographia* (Book 8) employed mathematical astronomy in a geographical context to draw a map of the world complete with meridians and parallels. According to the Alexandrian, geography treats "the form and size of the earth and its position in relation to the firmament, enabling the geographer to calculate the extent and location of its known parts; he also needs to know the latitude of individual parts in relation to the heavenly spheres."[15] This Ptolemaic tradition could very well have been one source of merging the goals of cosmography with those of astronomy and requiring a consequent emphasis on mathematical techniques.

The most likely source of Kepler's cosmographical knowledge was Peter Apian's *Cosmographicus Liber* (1524). Kepler never referred to Apian's work explicitly, but he almost certainly knew of the cosmographic tradition by 1596. In his annotations to the second edition of the *Mysterium* (1621), Kepler referred to the contents of this tradition, as he complained that his book had been misunderstood by booksellers who classified it under geography. Kepler explained that his goal had been to treat the structure of the world system rather than geography properly speaking.[16]

Why then did he choose *cosmographicum* in 1596? The immediate reason lies in his understanding that *cosmographicum* indicated physical descriptions and causes, as is evident in his student disputation. In addition, if Kepler had entitled his work *Mysterium Astronomicum*, it would probably have been interpreted only in a mathematical context, and the physical interpretation that he so greatly sought would have been lost on his audience. The cosmographic tradition, in modified form, allowed Kepler to locate his first book in a tradition of topics which was explicitly physical and which had been well established prior to Copernicus.[17] Copernicus's heliocentrism provided Kepler with the best available system in astronomy, one that had the additional benefit of explaining how the world system could be physically true.

Kepler possibly also chose *cosmographicum* for its theological connotations, for he was well versed in classical Greek and was no doubt aware of the etymological significance of the term.[18] In his *Dedicatory Letter* to Frederick, Kepler argues for the sublimity of contemplating

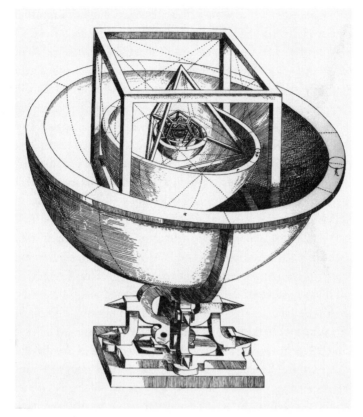

Johannes Kepler's diagram of the Platonic solids as explaining the distances of the planets. Courtesy of the Notre Dame Library.

"this most glorious temple of God" [the universe] in which "God [is] to be celebrated, venerated, and admired in true worship."[19] Kepler weaves together quotations from several Psalms to show that God has written his own grandeur in the fabric of the heavens. The book of Scripture issues a clarion call to open the pages of nature to find the "clearer voice" of God of which the Psalmist spoke.[20] In the *Preface to the Reader* Kepler explains how he concluded that the five regular solids provided the framework for the spacing of the planets. He contends that the "pages of nature" that he sets out to read are written by the divine scribe in the language of geometry. Kepler must plumb the depths of the divine will in the cosmos by using the appropriate tools of interpretation.

Mysterium proves to be more difficult to explain because there was no precedent in astronomy or any related natural discipline for choosing such a term. In the annotations to the second edition, Kepler emphasized the novelty of his polyhedral theory as one reason for his use of *Mysterium*. It was a secret that had never seen the light of day: "I have employed the term *mysterium* as a secret [*pro arcano*] and I have sold this discovery as such; indeed I have never read it in any philosophical book whatsoever."[21] Was the novelty of the ideas the reason he chose *mysterium* in 1596?[22] The polyhedral theory was certainly novel, but the word *mysterium* also carried a much richer connotation provided by theology, one that was appropriate for Kepler's self-imposed task. With two and a half years of theological study at Tübingen, Kepler had to be aware of the background of the modifications of Greek terms in the development of Christian dogma. The Pauline use of *mysterion* encompassed a spiritual truth or reality that was expressed through physical or material means, as when Paul speaks of marriage as a reflection of the relationship between Christ and the church.[23] Patristic theology developed this usage by speaking of the sacraments (Baptism, Eucharist) as the divine mysteries, visible and physical expressions of the divine presence.[24] Later in life Kepler was to differ with Lutheran authorities over the doctrine of the real presence of Christ in the Eucharist, an indication that he was well versed in the history of this doctrine and the sacramental use of *mysterium* in a theological context.[25]

I suggest that this theological background played a role in Kepler's choice of terminology. We know that Kepler was deeply concerned about theological issues during the 1590s, from his letter to Maestlin stating his desire to become a theologian. He intended to include in the *Mysterium* a discussion of the compatibility of Scripture and Copernicanism that the faculty senate of Tübingen advised him to drop while they unanimously approved its publication. We can certainly say that a theological significance for Kepler's choice of *mysterium* comports well with his goal of displaying the glory of God in his work. The two terms, *Mysterium Cosmographicum*, taken together, reveal Kepler's orientation to reading both the Bible and the heavens. God can be known in both books, but each has its own language.

If Kepler adopted the Melanchthonian emphasis on *sola scriptura* as a student at Tübingen, he probably believed that a priority should be given to Scripture for the purposes of theology. This may explain something of the disappointment he felt when called away to teach mathematics in Graz. His later joyful announcement that God could be known and celebrated in his astronomy meant that the *Mysterium Cosmographicum* represented both a personal fulfillment, as well as the commencement of a fruitful career that would lay bare the secrets of the universe in which the Divine Creator embodied his presence and plan.

In one of his last works (*Harmonice Mundi*, 1619) Kepler represented his own astronomy as an act of interpretation analogous to exegesis of the Bible. His exegesis of the universe was offered to God as an act of worship. Kepler constantly gave God credit for the success of his work, both in its discovery and in its reception.[26] He conceived of his discoveries as offerings to God because he believed that displaying the Creator's wisdom was a far greater sacrifice than offering bullocks. Quoting Galen and Plato on the necessity of invoking divine aid for investigating the cosmos, Kepler viewed his perfecting of Ptolemy's *Harmonics* as plundering the Egyptians. Alluding to the biblical Exodus, Kepler saw himself as "stealing the golden vessels of the Egyptians in order to build of them a temple for my God, far from the territory of Egypt."[27] It probably did not escape Kepler's notice that Ptolemy had resided in Alexandria. He likely intended this allusion in both literal and figurative senses. Christian authors had long appealed to the Exodus as a justification for incorporating pagan knowledge, but Kepler could now employ this language with a power that made it even more effective.[28] Ptolemy, the Egyptian, had provided the vessels for the astronomical Moses to reshape them into a temple for the true God. What pagan antiquity had begun, Kepler would bring to perfection.

Kepler conceived his task on the basis of a theological vision in a deeper sense than simply using his science to glorify God; it also shaped his view of what was necessary and progressive in astronomy. As we will see later, Kepler laid considerable stress on the celestial harmonies as the foundation of harmony in human society. He regarded celestial unity as the greatest expression of God's wisdom. Truth could not be truth if it were not unified. Consequently, in the *Astronomia Nova* (1609), Kepler divided the three major astronomies (Ptolemaic, Copernican, Tychonic) into two categories, depending on whether the planets were treated in a unified manner or not. The Ptolemaic system was inferior because it treated "individual planets separately and assigned the causes of motion to their own individual orbs."[29] Both the Copernican and Tychonic systems had the greater merit of asserting a common cause—the sun—for the motion of the five planets. From Kepler's vantage point, Tycho's partial heliocentrism is not simply a compromise between an ancient Ptolemy and a modern Copernicus, but a progressive step toward a unified description of the system, a direction that Kepler thought must necessarily lead to truth. For Kepler, any system is preferable that deduces planetary motions from a common cause.

Kepler then understood his astronomy explicitly in terms of Christian theology, a background that shaped his view of the goal of science (a unified truth of the physical universe) and its ultimate purpose (to glorify God). Kepler firmly believed in the integrity of both books of God, Scripture and nature, but he was always careful to distinguish the different languages in which each was written. Kepler's belief raises questions of how these books were related in his mind and how each should be interpreted.

KEPLER ON READING THE SCRIPTURES

Kepler had a complex view of the Bible, of how it should be interpreted and how it relates to nature. His quotations from and allusions to the Bible are found in a variety of contexts that provide windows on how he viewed relations between biblical authority and inquiry into nature. The notion that Kepler viewed the Bible primarily as a moral and not as a scientific authority suffers from oversimplification and is drawn from a limited number of his pronouncements on the subject of biblical interpretation.[30] He did indeed regard the subject matters and methods of astronomy and theology as quite distinct, but this distinction was acknowledged on all sides in Lutheran circles; he was not unique. He differed with Melanchthon's approach to the problem of Scripture and Copernican astronomy, but even this difference cannot be fairly characterized as an attitude of biblical irrelevance with respect to science. Kepler's approach to the Bible is properly illumined only when we see him as a combined product of his humanist, theological, and scientific training. In all arenas, Kepler displays an undeniable reliance on his past, and an unmistakable independence that came from his acute critical faculties.

In his theological training at Tübingen, Kepler no doubt learned the standard Lutheran view of the authority of the Bible and, like his teachers and confreres, he viewed the Scriptures as the source of truth in theological matters. The entirety of the Christian faith could be derived from them, a faith that was evident in the purity that characterized the early centuries of Christianity.[31] Several features of the Lutheran (and generally Protestant) view of the Bible become relevant for understanding Kepler's own views: unity, sufficiency, and clarity. As traditional Christian theology asserted, the Lutherans held that the parts of the Bible, in a complementary fashion, constituted a unity. No contradictions in the Bible could be admitted, an assumption that exegetes emphasized in following Augustine's dictum which said that Scripture must be compared with Scripture to resolve apparent conflicts between various texts.[32] Lutherans also maintained the sufficiency of Scripture for the purposes of faith and life, a belief that developed in special opposition to the Roman church's belief in the authority of oral tradition. The Lutherans insisted on their continuity with the early church, but they also maintained that this faith ultimately rested on the Scriptures alone.[33] The third feature of Lutheran teaching was the clarity of Scripture, an attribute that needed special emphasis because it formed the foundation of their hermeneutics. The Protestants were especially vulnerable to the Catholic argument that their internal divisions proved the insufficiency of relying on the Scriptures alone in formulating proper Christian theology, and Kepler himself was deeply troubled by the plurality of confessions among the Protestants. The perspicuity of Scripture was part of the Protestants' answer to the Catholic insistence on an authoritative magisterium in interpreting the meaning of Scripture.[34] This doctrine proved, however, more difficult to apply in practice than to formulate in theory. The Protestants acknowledged that not all texts of the Bible were equally clear, but for them this simply implied the need to follow the Augustinian guideline of comparing Scripture with Scripture. To arrive at proper interpretations, one should employ clearer texts to elucidate more obscure ones. More importantly, the clarity of Scripture was limited to those things that were deemed necessary for salvation, for morals, and for the life of the church.[35]

While Kepler regarded the Bible as an absolute and inviolable authority, he was also keenly aware that it was a collection of books from ancient history. For that reason, his reading of the Bible cannot be divorced from his reading of ancient texts in general, for he was well-versed in humanist techniques of exegesis and criticism, and on more than one occasion he was forced to engage in minute textual analysis of the kind expected of an experienced philologist.[36] Kepler's work on the history of astronomy and the scientific interpretation of classical texts (e.g., Lucan, Homer) reveals a man capable of and committed to departure from standard modes of interpretation when he thought it necessary. We saw in the previous chapter Tycho's view of ancient astronomy as arising among the biblical patriarchs and prophets, perfected among the ancients, and subsequently lost. This version was close to the predominant mode of interpretation that persisted well into the seventeenth century.[37]

Kepler's view differed significantly. The man who ventured to proffer an *Astronomia Nova* also willingly advocated a *Historia (astronomiae) Nova* because he saw the ancient and venerable discipline not as a *prisca theologia* to be revived but as an arduous task to be completed. In his history of astronomy, the biblical figures played no role whatsoever. The Greeks, rather, heirs to some Babylonian observations, systematized the observations under Hipparchus and Ptolemy, who together perfected this science and offered the first complete tables. Kepler proceeded to argue that the moderns (Regiomontanus, Reinhold, Tycho) had made the greatest contributions. Kepler's version in fact was not unprecedented, for a similar account had already been offered by Pico della Mirandola earlier.[38]

Kepler's treatment of the history of astronomy reveals one who devoted great effort to discern the intention of the ancient authors and the substance of the actual text, an attitude even more clearly seen in his hesitancy to read classical texts as containing scientific information. In 1599, Kepler's old teacher Michael Maestlin wrote him to request help in a task put to Maestlin by the eminent German classicist Martin Crusius. Crusius interpreted the encounters with the gods in Homer as planetary conjunctions and wanted Maestlin's help in calculating them.

As strange as such a procedure may seem today, we must remember that it was not unusual in Kepler's day for ancient literature, including the Bible, to be treated as authoritative for any subject that might be mentioned. Crusius himself viewed Homer as a source of ethics, economics, and philosophy, as well as poetry. Kepler's response to Maestlin's request was cool. In short, he refused to treat Homer allegorically, as a poetic expression of astronomical information. One may be tempted to think that his refusal was due to his separation of astronomy and exegesis, but a well-entrenched refusal to read the Homeric epic allegorically had existed since late antiquity. Kepler need not appeal to any modern science. Ancient forms of humanism would already insure a proper historical reading of Homer.[39]

This method of reading ancient literature indicates Kepler's careful attention to historical context, including consideration of the initial audience for whom the literature was intended. His reluctance to treat the Homeric epic allegorically can be ascribed to his belief that the ancients themselves would probably not have read it so. His meticulous analysis of Lucan's *De bello civili* points to his sense of the importance of literary genre and context for understanding properly the meaning of the text. This orientation opens up his approach to the biblical texts.

Kepler never engaged in any extended discussion of biblical hermeneutics (method), but from his handling of certain texts it is evident that he believed the intention of the biblical authors determined the meaning of the text. Discerning that intention required the interpreter to make judgments about the subject matter and the scope of the text in question. How was this to be done? As was true of classical literature, one must consider the historical and literary contexts. The only reliable method demanded close attention to the style and structure of the text. Kepler's most extended discussion of biblical texts is found in the Introduction of the *Astronomia Nova*, where he ventures brief answers to objections against the motion of the earth.[40] He poses for himself five different objections: three from physics, two from theology. The entire section on objections is embedded in a larger one that is designed to show how Copernicus provides a better celestial physics than Brahe. The two theological objections, disproportionate in length, appeal to Scripture and the Fathers, a combination common in sixteenth-century theological debates. Kepler's treatment of the Fathers, or pious antiquity, reflects a common Lutheran view of patristic authority. He praises the personal piety of the ancient Fathers but claims they hold no authority in philosophical matters: "while in theology it is authority that carries the most weight, in philosophy it is reason."[41] This selective treatment of the Fathers mirrors the Lutheran view of the primacy of the Bible in theological matters. The Fathers, though venerable, are nevertheless fallible and must not be taken *carte blanche*.

The length of Kepler's handling of the biblical texts exceeds by far his treatment of the Fathers. The texts Kepler cites can be divided into two types. The first type are those that appear to comment directly on the motion of the earth because they refer to natural phenomena. Of the major passages he discusses, four are of this type.[42] The second type are those employed by Kepler as a *reductio ad absurdum* to demonstrate that the biblical language cannot always be taken physically.[43] The latter type is important for understanding how Kepler argues for his interpretation of the former. This second category of texts containing well-known descriptions of nature are not interpreted in detail. Kepler uses them instead to argue that if such descriptions were taken literally, all inquiry into nature would cease or would lead to such absurd conclusions that anyone could recognize something amiss. For Kepler, these examples show that we must not attempt to read the Bible as if it were a work on *physica*.

This form of counterfactual reasoning is important to Kepler's purpose. He mainly wished to treat the four passages of the first type, three of which mention the sun moving or the earth at rest (Ps. 19; Josh. 10; Eccles. 1). Kepler frames the discussion by first arguing that ordinary language is tied to the senses and is not to be judged by philosophical standards. Even astronomers, when speaking outside the context of their work, retain the common modes of speech. No one would argue that this everyday language is inappropriate, because the astronomers are not attempting to speak in a philosophical manner at that moment. Then Kepler attempts to include the Bible in this everyday use of language. The intention of the biblical writers was not astronomical, and therefore there is no reason to think that they intended their words to be taken in an excessively literal fashion. The three passages that provided the standard objections are treated first, and in every case Kepler's method argues that the writers had a loftier purpose than to describe physical nature. He insists that they were not teaching astronomical lessons unknown to common people, but that they were recalling the obvious though forgotten lessons to be learned from nature, lessons of God's dealings with human beings.

Kepler could have contented himself with answering objections from Scripture regarding the motion of the earth, but he goes further to dispel any doubts about the intention of Scripture. He gives an extended treatment of Psalm 104 because it contains what many thought was a disputation on *physica*. Kepler maintains that the Psalmist's intention was doxological, not physical.

Speculation about physical causes is the farthest thing from the Psalmist's mind. The entire meaning of the Psalm lies in the greatness of God who made everything. He [the Psalmist] sings a hymn to God the Creator.[44] How did Kepler know that the Psalm was not about *physica*? Like any biblical exegete trained as a humanist, Kepler looked for connections between this text and other biblical texts. He concluded that Psalm 104 was in fact a commentary on the hexaemeron of Genesis chapter 1. Kepler repeated the well-recognized parallel structures of the first three and the last three days of Genesis 1, represented schematically in the diagram below.

Parallel Structures of Genesis 1

Form	*Filling*
Day 1	Day 4
Light Created	Light Bearers
Day 2	Day 5
Heavens	Filling Heavens
Day 3	Day 6
Earth	Filling Earth

This poetical structure corresponds to the six structural parts of Psalm 104.

We need to follow Kepler's method carefully because his exegesis of the psalm reflects his characteristic mode of reading. He analyzes the

Kepler's Analysis of Psalm 104 and Genesis 1

Genesis 1	Psalm 104
Day 1 Creation of Light	Section I Creator in Light v2
Day 2 Heavens Atmosphere	Section II Separation of Waters vv. 3–5
Day 3 Land Separated from Seas	Section III Earth vv. 6–19
Day 4 Luminaries Benefit of Man	Section IV Luminaries Benefit of Man vv. 20–25
Day 5 Filling of Earth	Section V Filling of Earth vv. 26–27
Day 6 Animals and Man	Section VI Animals Living v28

interrelationships within a text and infers a structural pattern which he then implicitly compares to a typical work on *physica* or astronomy. The structure of the psalm differs radically from that found in such works. Hence, he has some indication that the psalm is not a disputation on physical causes at all. Further, following the Augustinian dictum—stressed by Lutherans—of comparing Scripture with Scripture, Kepler notices the structural parallels between Genesis 1 and Psalm 104. Now he has found a reference point for the author's remarks; it is not nature directly that the psalmist is commenting on but an earlier biblical text. Kepler no doubt also thought of the genre into which the psalm fits. It would have been acknowledged on all sides that the psalms are praises of Israel, not historical narratives or cosmological treatises. Kepler then thought himself to be in a position to judge the author's intention. The very fabric of the text and its genre indicate a doxological, laudatory intention.[45]

Why did Kepler include this extended analysis of Psalm 104 in his Introduction to the *Astronomia Nova?* At one level, the answer is rather straightforward. Since the Tübingen faculty had advised him to drop any discussion of the compatibility of Copernicanism and Scripture from the *Mysterium,* he now—at a time when he no longer needed their direct support for publication—thought it necessary to dispel theological objections to heliocentrism. This explanation fails, however, to correspond to the length of his treatment since he could have simply invoked the general notion of accommodation and the non-astronomical character of Scripture without any detailed textual analysis. This would be entirely expected if he had been guided by an overriding separation of cosmology and theology. The length and detail of his discussion, however, can better be explained by considering the rhetorical function of his exposition.

Kepler carefully constructs the context of his exposition of the psalm by placing it last in the series of texts he treats.[46] Moreover, his entire treatment of scriptural objections is one of the longest—and next to last—of the five types of objections to the motion of the earth. It is evident from his phraseology that Kepler saved the hardest for last. He employs a form of argumentation *a majori ad minorem* to counter literalists' claims. If Kepler could show that Psalm 104 was not what was commonly believed, a physical disputation, he could then argue that other biblical texts—those not so manifestly about nature—were also not relevant to physical science and therefore could not be objections to terrestrial motion. To do this he invoked the internal unity of Scripture by comparing text with text. If Kepler could show that Psalm 104 had an internal structure that directly related to another text in the Bible, he could then argue that the psalmist had no intention of treating physical science but was rather commenting on God's earlier revelation.

A second argument was counterfactual. If the psalmist had wanted to deal with nature as an astronomer, then he would certainly have mentioned the five planets in the part of the psalm that is devoted to the celestial bodies (verses 20–25).[47] Here Kepler appeals to the undisputed belief that the Scriptures are the perfect and sufficient revelation of God. The psalmist would not have bypassed such an opportunity to reveal the glory of the Creator, but since he did not mention the planets, one can fairly infer that his intention was not that of an astronomer. To say that the psalmist failed to mention something so essential to his task would be to impugn the perfection of Scripture. Kepler does not say that the scriptural writers did not know the truth of nature, but only that they did not wish to comment on it. Thus, Kepler cleverly uses the Lutheran doctrine of Scripture to argue that the true intention of the Holy Spirit speaking in Scripture is not astronomical but doxological.

This exposition of Psalm 104 does not require that all of the Bible should be read as doxological. It could be argued that the psalm's literary genre indicates its special status and that other texts, such as Genesis 1, have distinctly historical and physical referents for their words. This would then mean that historical texts such as Joshua 10 with its implications about the motion of the earth should be taken as physical descriptions. Kepler's appeal to the notion of accommodated language was not limited to the psalms but rather extended to all Scripture, a judgment that was repeated many times during Kepler's life. As he was preparing the *Astronomia Nova*, Kepler stressed in a letter to Herwart von Hohenburg how thoroughly the Scriptures employed everyday language because even Genesis 1—an indisputably natural passage that could bear on science—should not be read as a natural-philosophical treatise.[48] The language of the creation account in Genesis was not to be seen as faulty or deficient; properly understood, the author simply adopted the common views of his audience in order to communicate more sublime truths.

Kepler's arguments on biblical interpretation should not be taken as an argument for an ultimate separation of scientific and theological truth because in Kepler's universe, as indicated earlier, the heavens were full of signs of divine wisdom and presence. His argument shows rather an interesting parallel in reading both the universe and the Bible. The divine intention must be discerned in both books in order to arrive at their proper meaning, and since intention can only be judged rightly by examining the language employed, the reader must know thoroughly the language of each book. For the heavens—whose language was mathematics—and for Scripture—whose characteristic mode of discourse was ordinary human language—interpretation crucially depended on knowing the relation between the vocabulary and the referents. Kepler had no doubt that geometrical constructs had a one-to-one correspondence with the physical referents in the universe, but the language of Scripture, like all human language, was imprecise and indirect. God in his wisdom had adopted a mode of discourse in the Bible that would not serve the purposes of astronomy but that admirably served the purposes of proclaiming the truths of salvation.[49]

Explaining Kepler's Universe

If nature and the Bible are written in two different languages, Kepler's universe is nevertheless filled with divine wisdom and presence. His spherical universe is finite, thereby distinguishing it from the infinite Creator, but this sphericity also manifests its origin because this shape, like no other in Kepler's mind, reflects the infinite and perfect trinitarian life of God. One may argue that in such a universe circles with uniform motion are the most aesthetically pleasing paths for planetary orbits. We know that Kepler—by far the strongest advocate of *a priori* necessity among the Copernicans—was long attached to this geometric tool. Yet Kepler also believed that the universe was a result of the divine will, and this belief required some modification of his strong necessitarian views. As he progressed in his attack on Mars, he did not yield to a theological voluntarism, but he did hesitate to ascribe his own preconceived notions to the Creator. After working out the elliptical orbits and the area law, he could still affirm the necessity of the cosmic structure, but he did so with a new openness to the geometric devices at God's disposal.

Kepler's universe can be explained by his realism, an integral part of which involves his beliefs about the relation of the Bible to nature. Kepler has long been seen as the first early modern who openly sought not only to save the phenomena of the heavens but to search for causes of planetary motion. The two traditions of mathematical astronomy and physical natures/causes of celestial phenomena existed in medieval science, but these two were kept quite separate. Chapter 3 argued that Tycho was seeking to overcome that barrier with his program for the restoration of astronomy, but it was Kepler who forthrightly demanded that the rigorous requirements of geometry and the physical causes of planetary motions be integrated. He knew that Ptolemy and Tycho could save the phenomena, but this was not enough. Kepler wanted the truth.

Why did Kepler adopt such a mathematical realism? Kepler scholars have often pondered this question, but they seem to lack a clear answer for several reasons. Some historians tend to think that only a historical answer could be given, and so they search for precedents in earlier science. Since there are very few to be found, they cannot arrive at a clear answer. A second reason lies in the tendency to separate the different aspects (astronomy, theology, and metaphysics) of Kepler's thought. These scholars have downplayed any possible linkage that might provide a key as to why Kepler took this realist approach. I will attempt to explain Kepler's realism by three complementary facets of his thought: his sense of moral obligation to the truth as correspondence, his sacramental view of nature (truth as embodied), and his view of truth as requiring unity.

Kepler's realism is explained in part by his sense of moral obligation to discover the truth of the universe as God made it. Hypotheses in the traditional sense of fictional constructs would not do, since they could indicate only the cleverness of the investigator, not the truth of the system. Kepler's repeated denials of searching for clever solutions and his affirmations of searching for God's glory aids our understanding of why he picked up the Ramian mantle and carried the prophetic torch of truth seeker.[50] Because he saw himself as an exegete of the Book of Nature, Kepler would consider himself disobedient to his Creator if he were to construct fanciful but false theories of the world. He often observed distortions of Scripture that were made for personal and factional purposes, abuses that warned him to seek honesty in his dealings with the Book of Nature.[51] He was always anxious to refute the charge of novelty-seeker or contentious spirit.[52]

Kepler was ready to adopt the challenge put forth by the French philosopher and logician Peter Ramus of constructing an astronomy without hypotheses (*sine hypothesibus*), a challenge that Tycho had rejected. Tycho and Kepler both were intent on finding the true system of the universe, but only Kepler was convinced of the absurdity of trying to prove the truth of nature with false hypotheses. Yet Kepler was also convinced that Ramus did not properly understand Copernicus, for Ramus had been misled by Osiander's *Preface* into thinking that Copernicus was using false hypotheses. Kepler's clever response to Ramus is designed to show that Copernicus was a true philosopher, that he sought truth about the heavens rather than simply saving the appearances.

Kepler's conception of scientific method is not limited to his view of hypotheses. As early as 1596, he indicated that metaphysical arguments were as, or more, powerful than mathematical ones. When he announced his belief in the motion of the earth, he promised to add to Copernicus's mathematical arguments his own physical, and better still, metaphysical arguments.[53] This high view of metaphysics also explains why he constantly invoked the icon of the Trinity as sphere. To him this was as powerful an argument for physical heliocentrism as any. God the Father is the source of all things and therefore must provide the moving force for the universe. The physical sun (not Copernicus's mean sun) both represents and embodies the paternal motive force for the entire system.

Kepler's moral sense of pursuing truth was conjoined with his sacramental view of nature. I suggested above that Kepler may have chosen *Mysterium* in the title of his first book because he conceived of the universe as a sacrament, a visible and tangible representation of the divine will and presence, an embodiment of God's essence that can be "seen" and known through the physical world. In 1599 Kepler wrote to Herwart von Hohenburg explicitly, "The world is the corporeal image of God, whereas the soul is the incorporeal, though created, image of God."[54] These indications, it turns out, are neither incidental nor occasional. Kepler's sacramental realism came to fullest expression in his often repeated analogy of the Trinity and the sphere. With the earliest expression in 1595 and the latest in 1619, the sphere icon functions as a central and controlling image throughout Kepler's life.[55] Although often noted in Kepler scholarship, the various instances of this image have remained unanalyzed—a particularly unfortunate lacuna since the variations emphasize different aspects. Perhaps the most frequently read version occurs in the *Mysterium*, where emphasis falls on the purely mathematical properties of the sphere. These properties remain constant throughout its many textual occurrences: the center represents God the Father, the surface of the sphere is God the Son, and the intermediate area is God the Holy Spirit. As a purely mathematical object the sphere-God relation might be thought to imply a simple correspondence between two otherwise dissimilar entities, an analogy imposed by a creative mind searching for connections. Certainly, the geometrical nature of the sphere is not incidental to Kepler's view of God, for he holds quantity to be an essential property of the world and a necessary attribute of God himself. However, Kepler transcends a purely mathematical similarity by holding to the physical reality of the sphere. One year before the publication of the *Mysterium*, Kepler connected the sphere with the entire universe, " Thus it is in the universe at rest: the fixed [stars], the sun, the heavens or intermediate ether: in the Trinity is the Son, the Father and the Spirit."[56] For Kepler the physical universe must be finite in space inasmuch as no more beautiful finite object exists than the sphere.

The full meaning of the sphere icon is not exhausted by its mathematical properties or physical reality. Kepler even more tightly connects the sphere with the divine trinitarian life. To understand how, it is important to ask why he identified the three members of the Trinity with the parts of the sphere that he did. Why did he not rather say that the area in the interior of the sphere was the Son and the surface was the Spirit? Or, why not let the center be the Son or the Spirit? Was Kepler's identification of the parts essential? Kepler responded with a resounding affirmation. God the Father must be the center, the origin of the sphere, because this origin, this "fount of Deity in the center," gives life and meaning to the whole object. Kepler considers the center point of the sphere to generate all the points on the surface. Mathematical generation corresponds to the generation of the Son inherent in the Trinity. God the Son must be the surface because every point on the surface is an image of the center as Christ is the image of the Father. The intermediate space must be God the Holy Spirit because any point in the interior of the sphere is equal to any other in relation to an appropriate distance between the center point and a point on the surface, and the Holy Spirit is fully equal to the Son and the Father. Although each part of the sphere is distinct from the other, there is such a mutual interdependence among the three that one part cannot be negated without all being destroyed, a claim that perfectly parallels the classic notion of God as necessarily trinitarian.

Some of Kepler's expressions of the sphere icon are replete with biblical allusions and traditional trinitarian phraseology. He speaks of "an absolute equality everywhere between point and surface, the sharpest union, the most beautiful harmonization, intertwining, relation, proportion, comparison." In the *Astronomia Pars Optica* and later in the *Epitome* Kepler speaks of the surface as "the image of an inner point" and "a way to find it" (i.e., the center), alluding to Christ as the image of God in Colossians 1:15 and the way to the Father in John 14:6. Similarly, in the latter work, when Kepler says, "whoever sees the surface, sees in it the center" he is undoubtedly alluding to Jesus' answer to Philip in John 14:9. He also refers to the surface as the "imprint" and "brilliance from it" (i.e., the center), alluding to the rather unique language of Hebrews 1:3. Kepler proceeds to describe the intermediate space of the sphere as "it measures and scrutinizes the depths of this figure," alluding to the Holy Spirit's activity in Romans 8:27.

This kind of language is only comprehensible in terms of certain aspects of the doctrine of the Trinity. Kepler's emphasis on the full equality of the Spirit draws on the results of the church's struggle in the *filioque* controversies. Kepler reveals his belief in the double procession of the Spirit—in keeping with the Western church's formulation—when he affirms that "the intermediate space results from a comparison of the center with the surface and proceeds from both."[57] He also employs the notion of the *circumincessio* (or *perichoresis*) when he speaks of the intertwining of the parts of the sphere or of the mutual love among members of the Trinity. The classic doctrine emphasizes that the divine life of the Trinity is not static but dynamic, a self-giving love and mutual indwelling of each divine person in the other. The realization of how deep is the sphere-universe-God connection in Kepler's thought only underscores that his view of the trinitarian mystery was such an essential feature of his cosmology that to remove it from his thought would be to destroy its inherent unity. He sought to maintain balance in the unity and distinctness of the

persons and to emphasize how the universe reflects each one in particular ways.

Another aspect of a sacramental view of nature recognizes that the
instrument of divine presence is accommodated to human understanding. We noted in the discussion of Kepler's biblical interpretation
how accommodated language pervaded the biblical texts, but we failed
to ask why Kepler might have interpreted the Bible in this manner. He
clearly invoked this concept to fend off the opponents of terrestrial
mobility, but this is only the most obvious reason, one that hardly explains the depth of his thinking. For Kepler, accommodation characterized not only the language describing nature in the Bible; it also guided
God's method of operation in the construction of the universe. Kepler
employed the verb *accommodare* and its synonym *aptare* in the opening pages of the *Mysterium Cosmographicum* to emphasize the internal
conformity of God's creative work. God accommodated the number,
proportions, and system of planetary motions to their nature when he
created the world.[58] This celestial harmony prompted the underlying
problem of the entire work: "since God adapted the requisite motion to
the distances of the orbs, surely he must have accommodated the distances themselves to some requisite measure."[59] What was this something (*ad alicuius rei praescriptum*) to which God adapted the creation?

As Kepler himself tells us, his revelatory moment came during a
lecture on 19 July 1595 when the five regular solids presented themselves
to him as the answer.[60] One of the most important concepts of the
Mysterium Cosmographicum, beyond that of the Platonic solids, came
to expression in chapter 22. Here Kepler anticipated his second law
(times-area law) by stating a relationship between speeds of the planets
and their distances from the sun. Kepler demonstrated that the
"planet's eccentric path is slow when higher, fast when lower" and that
this ratio was equivalent to Copernicus's epicycles and Ptolemy's
equants.[61] Recent investigations of his Mars manuscript reveal how
Kepler moved from the speed-distance relation in the *Mysterium* to the
times-area law of the *Astronomia Nova*. Kepler again employs the terminology of accommodation to explain the relation:

> Next, it is a physical fact that a planet proceeds slowly or quickly ac-
> cording to its approach to or departure from the sun. This is con-

firmed by comparing pairs of planets with each other. For the one that has the longer radius also has the longer delay [*mora*] over an equal space. Yet the difference is this, that when they are taken pair by pair the slowness is not entirely proportional to the circuit . . . but is deliberately adjusted to harmonic proportions. However, in one and the same [orbit] the slowness or delay is entirely proportional to the distance. What else could have been the reason for the accommodation of the eccentricity of the equant to the eccentricity of the eccentric?[62]

Kepler develops further several of the archetypal concepts already present in the *Mysterium*. Most important for our purposes is his belief that the speed of a planet has been "deliberately adjusted to harmonic proportions" (*est consilio accommodata ad harmonicas rationes*). His choice of the passive voice and *consilio* leaves little doubt as to whose deliberation he intended. The same Creator whose wisdom he vowed to reveal, deliberately chose to fit the universe to the harmonic motions and to shape the human intellect to understand these proportions.

The divine method of adapting to harmonic proportions parallels accommodation in the formation of Scripture. Since God accommodated the sacred writings to human language to be understood, ordinary language should not be discarded in favor of some technical language of astronomy because the Creator has chosen deliberately to form the sacred writings according to human linguistic capacity. In both cases, God has adapted his creative acts (or word) to the capacities of human minds. Kepler, then, had two compelling reasons to see the biblical language as accommodated. It was a well-embedded tradition of hermeneutics in Christian history, and his own understanding of the universe comported well with it. For him, all God's dealings with the universe are colored by accommodation because the cosmos was to be an instrument (sacrament) of celebration and adoration. Kepler's realism then is not merely the union of abstract mathematics and the physical universe. It involves rather the union of a necessarily quantified God—a specifically trinitarian God—with a universe that must thereby be necessarily quantified and trinitarian.

Kepler's sacramentalism did not demand, however, that he hold the kind of unity that is so evident in his thought. After all, Osiander and

Melanchthon believed that knowledge of God was possible through the created order and that God's nature and purpose could be discerned from nature, but they did not believe that mathematical astronomy was a key to this knowledge. Kepler clearly did. Without his Neo-Platonist commitment to the geometrical nature of the universe, Kepler would not have been a founder of modern science, an essential component of which was belief in a unified truth. From the *Mysterium* to *Harmonice Mundi* over twenty years later Kepler's underlying assumption remained that truth always led to unity, not disunity. This is most evident in his treatment of planetary theories in the *Astronomia Nova*, where one of the major achievements was a unification of solar theory with the models of the superior planets. Ptolemy, Copernicus, and Tycho all employed a bisected eccentricity in their models of the three superior planets (Mars, Jupiter, and Saturn), but their theory of the sun (Ptolemy, Tycho)—or alternatively the earth (Copernicus)—lacks this conceptual tool. Kepler's work extended a bisected eccentricity not only to earth's orbit but to the two inferior planets (Mercury, Venus) as well. In his mind, such a step of unification had mathematical grounds of harmony, but it was also buttressed by physical arguments.

As the title page proclaimed, the *Astronomia Nova* represented a search for causes, and so a significant portion of the book (chapters 32–39) was devoted to the beginnings of a physical hypothesis. Only a system founded on physical causes would suffice because the system is "so interconnected, involved and intertwined" that calculations fail on Ptolemy's and Tycho's system.[63] To achieve this, Kepler argued for a true sun as the center of planetary orbits, and he was not at all satisfied with Copernicus's mean sun. For this reason too, he thought highly of Tycho's system because it posited the sun as the center of the five planets, but Kepler went further by arguing that a modified Copernican system is superior to the Tychonic for the same reason—it makes the sun the real force for all six planets. This search for a unified treatment of planetary models based on physically real causes also explains his speculation in the *Mysterium*—and later his carefully worked out theory—that the speed of moving bodies is directly related to their distance from the source of motion, so that his underlying belief in the unified nature of truth contributed rather directly to his formulation of the area law.

It is tempting to leave Kepler's search for and attainment of a unified theory in the celestial regions, but to do so would leave unspecified
many of the connections he himself made. Perhaps most strikingly, he
unites the truths of nature and redemption in his first and last major
works. I noted above how Kepler argued against any method of seeing
astronomical truths in the Bible, and most Kepler scholarship has focused exclusively on these arguments. Consequently, the connections
between nature and redemption have either gone unnoticed or have
been treated as cultural ornamentation. However, contextual analysis
demands that we treat these passages as truly expressing his own
thought.

Recall that Kepler characterized the *Mysterium* as an inquiry into
physical cosmology without the expected proofs that should have been
in a book of astronomy. His polyhedral theory in fact served no standard astronomical function; it served only to show that the order and
spacing of the planets could not be otherwise, to demonstrate the *necessity* of the world structure according to the Copernican astronomy.
In chapter 4, Kepler wanted to explain why the earth is situated in the
middle of the planets with three bodies inside and three outside its
orbit. For this to be a problem at all, Kepler had to add the sun to the
planets (as well as ignoring the moon) since there are only two inferior
planets. Kepler's discussion is bracketed by the awareness that his readers may misjudge his intentions.[64] He offers an explanation of the
placement of the earth drawn from the "love of God for man [from
which] a great many of the causes of the features in the universe can be
deduced."[65] He assumes that his readers will agree that God does nothing in vain, and he ponders why God placed the earth in the middle of
the planets. Clearly, the "end of both the universe and the whole of creation" is the human race, and this arrangement serves "to provide evidence for the future tenant [of the earth] of the loving-kindness and
sympathy which God was to practice toward men, even as far as taking
his home among them as a friend."[66] Kepler argues that any other
arrangement would have been dismissed by the Creator as "an irregu-
larity of number lacking order." Kepler here expresses his conviction in
the unified purposes of God so that when the Creator set about to form
the world, he did it not only with humans in general in view, but
specifically with a view to the incarnation of the Son of God.

In the *Harmonice Mundi* Kepler ties the celestial harmonies to another aspect of redemption, the unity of the church. Here one senses the personal struggles Kepler faced with Lutheran and Catholic authorities as well as his intense dislike for acrimonious theological condemnations that he thought characteristic of those who would distort God's truth for personal ends. In the middle of his chapter on the origin of the eccentricities (book 5, chap. 9), Kepler breaks out in a prayer that calls for the universal church to display a new harmony, one modeled on the celestial harmonies.[67] This prayer, replete as it is with biblical allusions, has several features that make it more than incidental to his cosmology.[68] Kepler calls attention to the perfection of God's works in the heavens as a motivation for humans to emulate God by a perfect moral life. This moral life, expressed in deeds of love, will bring about the unity of the church on earth. As a priest of nature, Kepler has laid bare the harmonies of the heavens; it now remains for the church to fulfill its calling by bringing unity to mankind. At the same time, Kepler recognizes major obstacles to this unity. The devious and self-serving contentions of theological disputants will only foster "dissensions" and "sects." These are "works of the flesh" that are contrary to the purpose of God. Because of these obstacles, Kepler acknowledged the need for the help of the Holy Spirit as he also acknowledged God's help in his astronomical work. The most prominent feature of Kepler's prayer is his use of the high priestly prayer of Jesus in John 17. He models his own prayer on the Johannine one by beginning with "Holy Father" and by asking for the unity that comes from the mutual love in the Trinity. This appeal to the Trinity only confirms that Kepler envisioned both the celestial and the terrestrial worlds as a unified whole because they both were embodiments of the essence of God.

Kepler is nothing if not a subtle and complex thinker, so any attempt to summarize his thought runs the risk of inaccuracy. Nevertheless, some important conclusions can be stated. Studies of Kepler's theology and science have generally asked how the one influenced the other, with a discernible tendency for how the latter shaped the former. However, there is an inherent difficulty in framing the question in terms of disciplines, when Kepler himself sometimes did not make rigid distinctions. More perhaps than any other early modern astronomer, Kepler saw himself as an exegete of nature, a theologian whose task lay in opening up the Book of Nature. As is the case with many early moderns, the Christian religion motivated him in his work, but his own view transcends motivation and involves his conviction that the universe was an instrument of divine knowledge. Kepler's special mix of religion, metaphysics, and astronomy indicates the need for historians to distinguish between theology as a formal discipline and theology as a set of beliefs. Kepler rejected any necessity to confirm his own astronomical or cosmological conclusions from the discipline of theology. His arguments for separation were directed against theologians who overstepped their rightful boundaries. Nevertheless, these arguments do not support a wholesale divorce of cosmology from religious beliefs, since Kepler, more than any other prominent Copernican, had a forthrightly trinitarian view of the cosmos. Theology as a formal discipline was distinct from astronomy in his mind, but theology as a set of beliefs provided the metaphysical grounding for the true cosmology.

CHAPTER FIVE

Copernican Cosmology, Cartesianism, and the Bible in the Netherlands

*T*he influence of Lutheranism in western Europe was rivaled by the only other major force in the continental Protestant movement, Calvinism. Although many ecclesiastical notaries of various countries adopted or leaned toward Calvinism, the acknowledged center of the Reformed churches was the city of Geneva, largely because of the theological leadership of John Calvin. Calvin had arrived in Geneva in 1536 after publishing one of the most comprehensive apologies for the Reformation up to that time, *The Institutes of the Christian Religion* (1536, 1st ed.). By the year of his death, 1564, the Lutheran and Reformed movements, each bolstered by a host of its own confessions, had well-entrenched theological differences that were only deepened as the century wore on. It was the Calvinist Reformation that took hold in the Low Countries and exerted an influence that far exceeded the number of its adherents, so that by the end of the sixteenth century the Reformed faith had gained a foothold throughout the seven northern provinces. Although the Low Countries have always been at least 50 percent Catholic, the Calvinist Reformation found fertile soil on which to grow and made a much greater impact there than in Germany or France. Yet the specific character of the young Dutch Republic shaped the reception and advancement of the Reformed faith in ways very different from those in Geneva or in Hugenot France. The celebrated openness of the Netherlands, which allowed both a Catholic humanist such as Desiderius Erasmus (1467–1536) and an ardent Anabaptist like Menno Simons (1496–1561), also permitted an environment in which Reformed theology—now the new orthodoxy—could embrace Cartesian mechanism with few repercussions for theology. As we will observe, the openness of Dutch society and the refusal to endorse any national church were to have an impact on the reception of Copernicanism and Cartesianism in the seventeenth century.

In this chapter I explore the theological reception of Copernicanism in the low lands by the sea, a development that was to bind the fates of Cartesianism and Copernicanism together in the minds of advocates and opponents. Reformed theology did not in any way predetermine the outcome of the Copernican debates, but it did provide a distinctive context for debate, one that would reflect both the inner dynamics of Reformed thinking and also the special features of Dutch culture.[1]

THE EARLY RECEPTION OF COPERNICANISM IN THE NETHERLANDS

Copernican astronomy found little or no opposition in the Netherlands during the sixteenth century, but it also had very little exposure. An important early advocate was Gemma Frisius (1508–55) who was born at Dokkum in Friesland in 1508 and later became professor of mathematics and medicine at Louvain after taking the Master of Arts in 1528.[2] Frisius made several important contributions to mathematics and cosmography (geography), including a little book on triangulation that was appended to his 1533 edition of Peter Apian's *Cosmographicus Liber*. He also learned of the Copernican system from Johannes Dantiscus, who was the Polish diplomat in Louvain during Frisius's tenure there, although Frisius learned the details of Copernicus's theories only through Rheticus's *Narratio Prima* (1541). By 1545 Frisius had studied *De Revolutionibus* and cited it in his work on the astrolabe (*De astrolabio catholico et geometrico liber*), but his use of Copernican tables had to await his 1556 edition of *De astrolabio catholico*. In Rheticus's mind, Frisius rivaled Reinhold as Copernicus's successor when Rheticus spoke of the Dutch astronomer in 1550 as "another Copernicus during our time."[3]

In 1555 Frisius wrote an *Epistola* for the *Ephemerides Novae* of Johannes Stadius in which he lauded the superiority of the Copernican system for several reasons, not the least of which were its more accurate tables and planetary positions.[4] For Frisius, the science of astronomy was in flux, a state that required careful daily observations and tight geometrical demonstrations. Copernicus's achievement signaled the beginning of genuine progress (*initium progressumque*) over the extensive mistakes of the ancients (e.g., errors in Mercury's orbit).[5] And although the motion of the earth appeared absurd, Copernicus's system did not suffer from the same arbitrariness as Ptolemy's. Copernicus at least could demonstrate a necessary reason for many features that remained contingent in the Ptolemaic system.[6] It seems that the Louvain astronomer held to the traditional distinction between astronomy and physical cosmology characteristic of that generation of Copernican readers, but the beginnings of questions about the real motions of the heavens and their causes is also evident.[7]

The first decades of the seventeenth century were dominated by two of the most important figures in Dutch science, Isaac Beeckman and Simon Stevin. Beeckman, who felt a strong connection with artisans and craftsmen, became one of the first to articulate the classical principles of motion that would be independently formulated by Galileo and perfected by Newton. Although Beeckman never unmistakably came down on the side of Copernicus, his non-Aristotelian dynamics strongly favored a universe that did not have the earth as its center. More importantly, Beeckman developed a mechanistic worldview that was independent of Descartes at first and was later shaped by the association between the two. A strong Calvinist who knew Philipp Lansbergen and other Reformed theologians, Beeckman was no doubt aware of the theological questions revolving around the motion of the earth. It was his student Martin Hortensius who translated Lansbergen's 1629 *Bedenckingen Op den Dagelijckschen, ende Iaerlijkschen loop van den Aerdt-kloot* into Latin. In the final analysis, Beeckman never pronounced on the issue, but his scattered discussions of natural philosophy and theology did set the stage for the strong separation arguments offered by Calvinists later in the century. For Beeckman, the methods and subject matters of natural philosophy and theology were distinct and proceeded in different directions, the former "from wonder to no wonder" and the latter, in reading Scripture, "from no wonder to wonder." Philosophy led to knowledge of nature without ambiguities; theology led to the praise of the Creator who was unmistakably beyond comprehension.[8]

Simon Stevin, engineer and technical advisor to Prince Maurice of Nassau, was to introduce or at least codify much of the scientific vocabulary of the Dutch language (e.g., *wiskunde* = mathematics) in writings that ranged from memoirs on the excellence of Dutch to the principles of hydrostatics. In his theoretical studies, Stevin wrote on mathematics and dialectics, and in his single work on astronomy/cosmology he endorsed the Copernican system in his 1605 *Van den hemelloop* (on the heavenly motions).[9] Working mainly from the ephemerides of Johannes Stadius, Stevin argued for Copernicanism on grounds of simplicity and naturalness. In the third part of this work, when he treats the earth's motion, Stevin adopts a realist viewpoint in his belief that the planets are carried on spheres that interact with one another in a holistic dynamical system. Such a system was conceived in terms of Gilbert's magnetism, in which each orbit exerted an influence on others.[10] These early endorsements of the Copernican system paved the way for the theological discussions that began to emerge in the 1630s. By that time, too, Kepler's work and advocacy of Copernicus was widely known. Before examining the controversies of the 1630s, it is essential to understand how the Reformed faith shaped the ecclesiastical world of the northern Netherlands.[11]

THE THEOLOGICAL CONTEXT OF THE SEVENTEENTH-CENTURY NETHERLANDS

The development of Reformed theology in the Netherlands and its interaction with cosmological and natural-philosophical thought would have been impossible apart from the seminal influence of the Genevan Reformer John Calvin (1509–64) whose theology and ideals were transmitted very early to the low lands. Two aspects of Calvin's thought prove to be relevant here: his views of knowledge, with its implications for the relation of Scripture to astronomy, and his view of the church Fathers.

Understanding Calvin's approach to Scripture vis-à-vis natural philosophy must be set within the context of his view of the liberal arts.[12] Calvin viewed knowledge derived from disciplines other than theology positively and warned against rejecting it simply because it came from sources other than Scripture; he viewed all true knowledge as a gift from God. Among philosophers especially may be found "exquisite researches and skillful description of nature." Calvin praised those who "tasted the liberal arts [and who] penetrate with their aid far more deeply into the secrets of divine wisdom."[13] From his earliest work in Paris on Seneca's *De Clementia*, Calvin displays a strong humanist background that understood knowledge as a worthy pursuit. Such views characterized the mature Calvin as well. In 1550 he wrote a work entitled *De Scandalis* in which he sought to clarify some of the objections against the Reformation that had arisen from Renaissance humanist circles in France. Calvin spoke of the revival of learning in positive terms, as it had advanced knowledge in areas such as jurisprudence, medicine, astronomy, rhetoric and similar (*eiusmodi*) arts.[14] Knowledge of God was the foundation of all knowledge (*principium verae scientiae est dei cognitio*) and thus man's pursuit of knowledge is both honorable and must be guided by divine wisdom—two reasons why Calvin called for a *christiana philosophia* as a distinctly Christian way of viewing knowledge in contrast to the skeptical humanists' conception of knowledge.[15]

Calvin also viewed favorably knowledge derived from natural philosophy and occasionally drew upon it in explaining some biblical word or phrase.[16] An example is his attempt to explain why Genesis 1:16 speaks of the sun and moon as the greater and lesser lights when astronomers know that Saturn is greater than the moon.[17] Calvin explained the language of the Scriptures as being directed toward the human observer and clearly distinguished the purposes of astronomy and the Scriptures.[18] The purpose of the Scriptures was not primarily to teach cosmology but to instruct humans in their salvation. In fact, Calvin seemed anxious at times to fend off criticism of the Scriptures as having an inferior cosmology. His method of accomplishing his defense was to note that the Scriptures are designed to instruct the common man; "astronomy and other recondite arts" must be learned elsewhere:

The Holy Spirit had no intention to teach astronomy; and in proposing instruction meant to be common to the simplest and most uneducated persons, he made use by Moses and the other prophets of popular language, that none might shelter himself under the pretext of obscurity, as we will see men sometimes very readily pretend an incapacity to understand, when anything deep and recondite is submitted to their notice. Accordingly . . . the Holy Spirit would rather speak childishly than unintelligibly to the humble and unlearned.[19]

Calvin viewed all natural phenomena as a testimony of the providential care of the Creator over his creation, something that would be true no matter what particular science one might hold. Calvin's *Commentaries* show one who, as an interpreter of the biblical text, felt constrained to reflect what the text of Scripture was asserting. His exegetical task was not to debate issues of cosmology but to expound the meaning of the text *in the language* of the text. If the Scriptures were given in common human language, then the interpreter's task was also to expound the meaning of the text without reference to theoretical inquiries. In this conception of the task there is a clear distinction between the tasks of natural philosophy and theology.

What then were Calvin's views of Copernicus? Did he, in a view made popular by Andrew White over one hundred years ago, invoke Scripture against Copernicus? Such a polemic would hardly be expected given Calvin's views sketched above. After twenty-five years of scholarly debate, it seems fair to conclude that Calvin never opposed Copernican ideas directly, and it is doubtful that he ever knew of Copernicus or his system.[20] The clearest reference to heliocentrism in his *Eighth Sermon on I Corinthians 10–11* seems to refer on analysis to other matters than heliocentrism. Kaiser argued that Calvin's reference to geodynamism refers to Cicero's *Academica* rather than to Copernicanism, and so he concurs with Rosen's original analysis of Calvin's lack of Copernican knowledge.[21] According to Kaiser, Calvin was committed to a geostatic view of the universe because of his belief that this constituted a "clear testimony to the particular providence of God."[22] The Reformer inherited a medieval cosmology based on Aristotle which accepted most of Aristotle's postulates, but his reason for believing in the immobility of the earth was more theological than natural-philosophical. An immobile earth was "a precious sign of God's care for his people even in a world of turmoil."[23] In speaking of the stopping of the sun in Joshua 10 Calvin remarked that the daily rising and setting of the sun does not happen "by a blind instinct of nature but that he himself [God], to renew our remembrance of his fatherly favor toward us, governs its course."[24] If Kaiser's analysis is correct, Calvin would then have been perhaps disturbed by geokineticism, but there is simply no clear evidence that he opposed Copernicanism.

Calvin focused more on theology and exegesis by leaving natural-philosophical debates to others, because he himself was not knowledgeable enough about those ideas to adjudicate their adequacy. But he clearly drew heavily on the hermeneutical tradition of accommodation to defend the Scriptures against the charge of having an inferior cosmology. His positive view of nontheological disciplines and the culturally conditioned nature of the Bible was to have a widespread influence among Reformed thinkers.

Calvin was a younger contemporary of Luther and other first-generation Reformers, and consequently his interpretations of the Bible and theological doctrines depended heavily on appeals to the historical precedent of the church Fathers, particularly Augustine. Calvin's dislike of novelty for its own sake is evident in a little-known but significant letter to Grynaeus:

> Since in this life we cannot hope to achieve a permanent agreement in our understanding of every passage of Scripture, however desirable that would be, we must be careful not to be carried away by the lust for something new, not to yield to the temptation to indulge in sharp polemic, not to be aroused to animosity or carried away by pride, but to do what is necessary and to depart from the opinion of earlier exegetes only when it is beneficial to do so.[25]

Calvin's humanist training required him to be widely read in the early church Fathers, something dramatically seen in his speech at the Lausanne disputation in October 1536. Calvin quietly listened to the proceedings of this large disputation for four days, refusing to speak even though he was urged to do so by Pierre Viret and Guillaume Farel. On the fifth day, a spokesman for the Roman church read a long speech in which he charged the Protestants with ignoring and despising the church Fathers in their doctrine of the Eucharist. Calvin rose to refute these charges, weaving quotations together while emphasizing that these quotations were only representative of the Fathers' wider views. His speech is especially significant in light of the Catholic charge that the Protestants ignored the tradition of the church in their doctrine of *Sola Scriptura*—the notion that the Scriptures were the sole basis of theological formulation. Calvin's answer appealed to the historical precedence of Cyprian and claimed that the Protestant doctrine of *Sola Scriptura* was in the tradition of the Fathers.[26]

In 1539 Calvin was relieved of his ministerial duties in Geneva and moved to Strasbourg where he became a pastor of a French-speaking church. Knowing of Calvin's absence, the Italian cardinal Jacobo Sadleto wrote a lengthy letter to the city fathers of Geneva in which he urged them to return to the mother church. Sadleto charged that the Reformers were schismatic and that they had forsaken the teachings of the ancient church. In his reply Calvin agreed that leaving the church was reprehensible but argued that the Reformers had not done this; they had only revived and reaffirmed what the holy Fathers had taught. On each theological issue Calvin argued, "all we have attempted has been to renew the ancient form of the church."[27] On one of the central theological issues between Catholics and Protestants, the Eucharist, Calvin sought to answer Sadleto's accusations by an appeal to Augustine.[28] Calvin's goal was *renovatio*, an attempt to lead the church back to its roots, a consciousness that guides him in his interpretation of the Bible and Christian theology. In his magnum opus Calvin repeatedly sought to answer the charge of novelty by appeal to the Fathers, among whom Augustine was "the best and most faithful witness of all antiquity."[29] Nor was Calvin exempt from using an allegorical hermeneutical method prominent among the early Fathers, though he at times spoke against it; he employed a limited use of it similar to Augustine's.[30]

In sum, Calvin's thought was deeply molded by his humanist background, both in his positive view of the liberal arts and his dependence on Christian and pagan antiquity. Perhaps it was that same background that led him to distinguish the methods and goals of scriptural interpretation from those of natural philosophy—natural philosophy viewed as a divine gift that could be profitably employed for the greater glory of God. He did not address himself to the Copernican problem for one of two reasons. He either did not know of it or he did not consider it within his province to address it. Since he distinguished between disciplines and their methods, he knew that astronomy had its methods and that biblical interpretation also had its own. The task of the biblical interpreter was not to reconcile the results of knowledge gained from other areas of inquiry with biblical knowledge. The purpose of the Bible was to confirm the life of the faithful, not to teach science, whether Aristotelian or any other. Calvin's extensive development of a method of accommodation suggested that any potential conflicts between nature and Scripture could be dealt with by a deeper analysis of the biblical text.

The influence of the Reformed religion was felt in the Netherlands very early in the sixteenth century, but its reception had a mixed history due to the political uncertainties of the times. Calvinism began to emerge as a major force in Dutch culture with the synod at Emden in 1571, a local event that was to have an impact across all the northern provinces. This synod and the churches associated with it represented a resurrection of the Dutch Reformation that had been all but annihilated in the early decades of the Reformed movement (ca. 1520–1540).[31] With the revolt against Philip II and Spanish rule, the particular towns and provinces became far more important in guiding the political and cultural vision of the Dutch. The consequent separation between various regions of the low lands engendered great diversity of belief and practices, but two events soon counteracted this tendency by bringing about cooperation. The most obvious was the Union of Utrecht negotiated by William of Orange and his brother John the Elder in 1579, a move that united the seven northern provinces. The second was the growing adherence to the Reformed religion among the inhabitants of the seven provinces in the Union. While the Reformed Church never was an official state church requiring membership of all inhabitants, there was a practical control of local affairs in the towns that required all regents to endorse the Reformed religion to varying degrees.[32]

During the years of the truce with Spain (1609–21) the northern provinces were filled with religious controversy, the most prominent of which was the dispute between the Leiden theologian Jacob Arminius and his colleague Francis Gomarus. Arminius advocated a softer version of predestination and atonement than had become standard among the Dutch Calvinists. Because Arminius and his "remonstrants" were in a decided minority, they often had to invoke a rhetoric of tolerance in order to gain a hearing among the Reformed ministers. The several attempts to reconcile the Arminian remonstrants and the strict Calvinists ("the precisians") were totally unsuccessful. Finally, the States-General called for a national synod to meet at Dordt in 1619. The synod, while excluding the remonstrants, issued the *Canons of the Synod of Dordt* that proved to be the definitive response to the remonstrants and the guiding documents of the Reformed churches for centuries. While the details of the differences between Arminius and his opponents need not be recounted here, the fallout from the synod dramatically altered the context of theological discussion because of the complete hegemony of the Calvinists. Yet the Reformed religion was flexible enough to allow for diversity of opinion on secondary matters. What constituted a "secondary" issue was itself a matter of dispute, but clearly natural philosophy was not as central to the ministry of preaching as issues decided by the Synod of Dordt. These developments produced at once an environment in which concern for theological accuracy and openness on natural philosophy coexisted.

PHILIPP LANSBERGEN AND THE REFORMED DEFENSE OF COPERNICANISM

One of the most explicitly theological defenses of Copernican cosmology came from the pen of the Reformed pastor Philipp Lansbergen in 1629 entitled *Bedenckingen Op den Dagelijckschen, ende Iaerlijkschen loop van den Aerdt-kloot* (considerations on the diurnal and annual motion of the earth) and was translated into Latin the following year by Martin Hortensius.[33] By the time of writing this treatise Lansbergen had spent a long life as a minister and had already written a major work on Copernican astronomy as well as printed sermons and a work on biblical chronology.[34] Lansbergen's motivation went beyond Kepler's desire to demonstrate the glory of God; his goal also included a specifically ministerial dimension. As he ended his treatise, his years of pastoral service were reflected in the comfort he hoped these considerations would bring to his readers. A hint of the doxology so characteristic of Reformed worship can be detected behind his words, "I heartily worship God through our Lord Jesus Christ who is everywhere praised and adored as God should be praised forever. Amen."[35] Astronomy as much as theology could be the instrument of offering praise to the Creator and comfort to human creatures.

Lansbergen viewed science as a pursuit of truth about nature, not simply saving the appearances. The Copernican theory was not just a convenient instrument for prediction; it was the only true system because it fit "nature, reason and truth" better than Ptolemy's model.[36] While Ptolemy sought to predict the planetary positions and motions, Copernicus went further and aimed at elucidating the causes behind those motions.[37] By contrasting the respective hypotheses, Lansbergen promised that the Copernican system would be established on the firmest foundation possible.[38] His rejection of Ptolemy was based on a strong sense of the orderliness of nature in which certain claims were deemed absurd because they contravened the regularity and rationality of nature. Real physical arguments for Copernicanism were compelling because the opposite Ptolemaic position implied physical impossibilities. The problem of diurnal motion is a case in point.

On Ptolemy's theory, the alternation of day and night was explained by the movement of the sphere of the fixed stars (the eighth sphere) which revolved every twenty-four hours. On the Copernican theory, of course, the rotation of the earth with respect to the sun served the same explanatory purpose. Lansbergen argued that attributing to the sphere of the fixed stars a motion fast enough to account for the alternation of day and night would make this largest sphere move the fastest. This is a "manifest absurdity" because the principle that guided nature was "the smaller the circle, the faster the motion."[39] Lansbergen concluded that on the Ptolemaic position the eighth sphere would have to transverse almost 643,980 German miles per second to fulfill the requirements of the assumed mathematical distances. No physical body could move so fast. By contrast, the Copernican system required the earth to rotate only about three German miles per minute. Surely, a system with a moving earth fit reason better, especially since the four elements of which the earth was made are more subject to

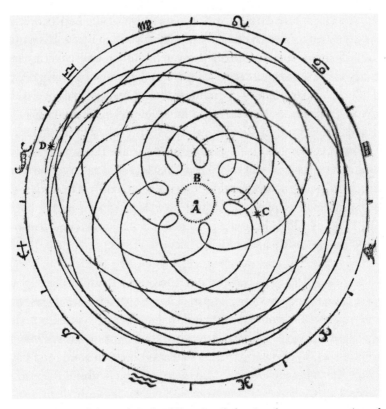

From Philipp Lansbergen's *Bedenckingen* (1629) showing the apparent motion of
Mars from 1580 to 1596. Lansbergen drew this diagram from Kepler to show the
elliptical orbit of Mars. Courtesy of the Bibliotheek van de Gemeente Universiteit
te Amsterdam.

change and decay than the eighth sphere.[40] The criteria for judging between competing world systems for Lansbergen clearly depended on
what fit best with a causal analysis of motion and reason in general.
The deeper foundation of this view of knowledge lay in the same confidence that Kepler expressed, that of a rationality of nature based on a
rational God.[41]

It was not enough, however, to deal with the astronomical reasons
for Copernican superiority; the theological issues also had to be addressed. We may infer that the theological question had become sufficiently troubling in the Netherlands because the famous printer of
Amsterdam, Willem Blaeu, encouraged Lansbergen to write a book in

Dutch dealing more directly with the question of the earth's motion; Lansbergen responded by dedicating his 1629 book to Blaeu. His writing in Dutch was no doubt directed toward church and governmental officials who were not sufficiently trained in the mathematical and theological disciplines but who were nevertheless responsible for making decisions that would affect both the churches and municipalities of the provinces. A tightening theological climate prevailed in the 1620s as a result of the Synod of Dordt's condemnation of the Remonstrants. This situation may have caused the likes of Blaeu and Lansbergen to worry that other divisive issues would incur the wrath of reactionaries who would make astronomical questions into theological dogmas. Such a fear would explain Lansbergen's strategy to distinguish between a knowledge of mathematics and the purpose of Holy Scripture:

It is absurd to judge the motion of the earth from Holy Scripture. Of course when the foundations of these questions are raised in astronomy and geometry, they are not to be answered from Holy Scripture. I agree with the Apostle that all Scripture is divinely inspired and useful for doctrine, rebuke, correction, and teaching in righteousness (2 Tim. 3:16). But I deny that it is useful for instruction in geometry and astronomy. The Spirit, as author of the Holy Scriptures, did not desire to hand down the foundations in either of these sciences [*ars*]. Those things that support the principles [of geometry and astronomy] are not properly learned from Holy Scripture but from those who are engaged in it. Thus, we read that Moses had learned them not from Scripture but from the Egyptians (Acts 7:12) and Daniel with his associates had received them from the instruction of the Babylonians (Dan. 1:4,10). So they miss the purpose of Scripture who see geometrical and astronomical questions as its norm.[42]

Hooykaas took statements like this as characteristic of Lansbergen's hermeneutical approach and as one more reflection of Rheticus's separation of scientific and theological issues.[43] I argued in chapter 2 that Hooykaas mistook the complexity of Rheticus's approach to the Bible and the relation between natural inquiry and theological dogma. Yet Lansbergen, even more keenly than Rheticus, strove to demonstrate the nonscientific character of biblical language by citing mathematical examples in Scripture that could not possibly have been intended to give mathematical knowledge or empirical truth.

In 1 Kings 7:23 and 2 Chronicles 4:2 the Scriptures refer to the ratio of the circumference of a circle to its diameter as 30/10 or 21/7. Archimedes showed, says Lansbergen, that the ratio is really 22/7. Scripture was not so much wrong as it was speaking in the manner of common speech of that day (*populari & recepto more*). Similarly, the church Fathers Augustine and Lactantius discussed whether the earth was spherical and whether there were antipodes. Lactantius denied both and derided anyone who held the contrary. Augustine, though not denying the sphericity of the earth, concluded that there were no antipodes because Moses does not mention them after describing the flood, and other scriptural texts seemed to confirm this as well (e.g. Ps. 24:2).[44] Later theologians, such as Virgilius, argued that the existence of antipodes would require another Christ to have died. The reasons and experience that astronomers have adduced demonstrate that these theologians were ill-advised to cite scriptural testimonies in behalf of their wrong opinions. Lansbergen read the Fathers as interpreting the Bible in too physical a manner. This manner of reading mistook the intention of the biblical authors whose goal was something else and who employed the common knowledge of their day to convey theological truths.[45]

Lansbergen believed that the same misguided view of Scripture stood behind purported scriptural objections to the motion of the earth. The Psalmist speaks of the earth being founded on the seas (Ps. 24:2), a natural way of speaking since the ocean appears more extensive than the land. But reason proves the opposite; the seas cannot exist by themselves but must rest on the earth.[46] Why has the Psalmist spoken in this manner? Lansbergen appealed to the same Origenist metaphor that Calvin invoked:

> For it is certain that the Almighty God fills heaven and earth. Although the Holy Spirit stammers [*balbutiens*] with us, he says that he often descends to us and then goes back from us because we perceive it better by human manner of speech [*humano more*]. So, when it asserts that the sun moves and the earth stands, it is nothing strange because they offer something to our eyes that is different from what is.[47]

This appeal to accommodation aided in fending off scriptural objections against diurnal motion, but it also indicates how extensive is Lansbergen's separation of astronomy from theology. The Bible has a pastoral purpose, and to invoke it as an astronomical or physical authority is to misunderstand its language, words that necessitate nonliteral interpretations. Lansbergen's reasoning follows the straight-line accommodationist argument: show that the difference between the known reality (the Almighty God fills heaven and earth) and the language of the text (the Spirit often descends to us and then goes back from us) demands a nonliteral reading and then apply the same method to the new controversy. To Reformed readers who may have been familiar with Calvin's appeal to the Alexandrian Father, Origen, such an argument would have had a strong appeal.

Most scholarly discussions, such as Hooykaas's, content themselves with this appeal to accommodation or to the argument for separation of natural philosophy and theology, but these strategies do not exhaust Lansbergen's use of the Scriptures in a cosmological context. A second major portion of his treatise delineated the true form of the heavens as consisting of three levels, the first made up of the sphere of the planets, from the sun to the sphere of Saturn, and the second the sphere of the fixed stars. The third and final heaven was the abode of the elect, invisible to the physical eye.[48] Quite naturally, most of Lansbergen's citations from Scripture relate to this third or invisible heaven in which a third *energeia*, called the Spirit of God in the Psalms, acts as the breath of God renovating the earth.[49] There are many types or figures of this heaven in Scripture, the most obvious being the tabernacle because it is divided into three distinct parts (outer court, holy place, holy of holies). The tabernacle was called a type in Scripture (Heb. 9:24), "so it follows that the other two parts of the tabernacle [i.e. outer court, holy place] were also types of the two heavens which are open to constant inspection [i.e., visible]."[50]

If the tabernacle in Scripture is a type of the third heaven, the first heaven (planetary system) also serves as a reflection of something theological. Thus, when Lansbergen sketches the "true form [*typus*] of the first heaven," he not only treats standard astronomical parameters such as apogee, perigee, and mean distance, he also adduces reasons for the discovered symmetry in the orderliness of the Creator.[51] He adopted the Keplerian analogy of the universe with the Holy Trinity but applied

a new twist to this metaphor. We saw in chapter 4 that Kepler had identified the sun with the Father, the Son with the outer sphere of the universe, and the Spirit with the intermediate space. Lansbergen employed
the image in an entirely different manner and limited its application to
the relation of the sun, the moon, and the air:

> God has placed a kind of likeness in the sun. As there are three who
> testify in heaven, the Father, the Son and the Holy Spirit (1 John 5:7),
> thus there are three in the first heaven to illumine the earth: the
> sun, the moon, and the air around the earth. And these three are
> one and the same light. The diversity of subjects indeed shows a
> Trinity: of the sun, moon, and air. But as the Father has light from
> himself, the Son from the Father, and the Holy Spirit from the
> Father and the Son, so light of the sun comes from itself, that of
> the moon from the sun, and that of the air from the sun and moon
> at the same time. So there are in fact three different lights but the
> light itself clearly shows a unity: the light of the moon is the sun's
> own light and the light of the air is that of the sun and moon to-
> gether. In fact, there is only one light and the same proceeds from
> one sun as its source.[52]

Like Kepler, Lansbergen underscored the connection between the sun
and the Father because they both serve as the single source for their respective systems, but he has clearly identified the respective parts differently, due perhaps to his emphasis on the light that proceeds from God
the Father rather than on the mathematical features of the universe. It
seems that for Lansbergen the mathematical symmetry of the planetary
system (the first heaven) was a necessary but not a prominent feature.
His peculiar identification of the parts that correspond to the individual
members of the Trinity reinforced the Reformed adoption, in keeping
with all Western Christendom, of the *filioque* clause in the Nicene
Creed, which affirmed the procession of the Spirit from the Father *and
the Son*. Lansbergen's belief in the theological dogma of the *filioque* and
his version of Copernican cosmology both rest on the notion of celestial influence from God through the sun down to the earth—especially
on that crown of creation, the human race.

 Therefore, it is not surprising to find Lansbergen drawing on
the macrocosm-microcosm analogy that we observed in Tycho Brahe's

cosmology, that between the human body and the planetary system. Lansbergen focused on the physical properties of the human body and the universe itself, the sun being central because it perfectly reflects the position of the human heart in the body. The three heavens correspond to the three main faculties of the human body, the vital (heart), the natural (liver), and the intellectual (brain).[53] To both the universe and the human body God has given the heat and warmth of the sun as a life-giving force.

The focus on human life is also seen in the prominent role that the eighth psalm played in Lansbergen's thinking about the universe, a position it had for centuries in Christian thought. The Psalmist's exclamations of the majesty of God and the creation, the orderliness of the heavens, and the privileged position of man all suggested to Lansbergen that the ordering of the planets could not be arbitrary. Contrary to the older notion among historians that Copernican astronomy dethroned man from his medieval pedestal, Lansbergen's argument *for* a Copernican arrangement rested precisely on the uniqueness that the human race held in God's ordering of the planetary spheres.[54] Drawing on the Psalmist's language, Lansbergen saw the middle position of the earth between the inferior and superior planets as an expression—not of numerical symmetry as Kepler did—but of God's beneficence in crowning man with glory and honor (cf. Ps. 8:5).[55] The same divine care for humanity of which the Psalmist spoke reverberates through Lansbergen's discussion of the meaning of diurnal and annual motion. The daily rotation of the earth makes man a recipient of and participant in the *life-giving power* of the sun and demonstrates the maternal care of God for the crown of his creation. Annual motion affords man the opportunity for inspection of the divine works and thereby evokes praise and adoration for the Author of nature.[56]

When Lansbergen argued for the Copernican system on the basis of reason and truth, his notion of reason was not that of later astronomers or of historians viewing the seventeenth century through the lenses of twentieth-century achievements. His "reason" appealed to a symmetry with contemporary medical knowledge and theological underpinnings. He did, of course, argue as strongly as anyone for the separation of astronomical and theological questions as well as for a limited role of the Bible in settling physical issues. But also, like Kepler, he depended on the Bible in a deeper manner by his emphasis on the universe as an emblem of God's being and as a structure designed for human honor and benefit. The Copernican system was the most natural and reasonable because it made mathematical, physical, and theological sense. For Lansbergen this implied not that heliocentrism took one more step away from the Bible, but that it provided yet another vehicle through which the Creator would be lauded and praised.

EXPOSING THE CALVINIST-COPERNICAN SYSTEM: LIBERT FROIMOND'S *ANTI-ARISTARCHUS*

No sooner had the Latin translation of Lansbergen's *Bedenckingen* appeared than a vociferous response emerged from the center of Catholic orthodoxy in the Low Countries, the University of Louvain.[57] Leading the attack was the recently appointed *Professor Ordinarius*, Libert Froimond, whose previous studies in astronomy, natural philosophy, and theology well equipped him for responding to all aspects of this issue. His critique was motivated by his fierce loyalty to the Roman Catholic Church and by his conviction that the decision to place Copernicanism on the Index (1616) must be defended against the heretics at all costs.

Scholars have sometimes referred to Froimond's pejorative characterization of that "Calvinist-Copernican system" to indicate his reactionary attitude toward anything novel, but too much should not be made of this designation because Froimond himself was not one to revert to older systems either in theology or in natural philosophy.[58] In the year that he received his doctorate in philosophy at Antwerpen, Froimond published *Saturnalitiae coenae, varitae somnio, sive peregrinatione caelesta* (1616), in which he rejected the Aristotelian doctrine of the four elements and the entire system of celestial spheres, expressing also Copernican sympathies in more than one place. Two years later, however, Froimond published his observations of the comet of 1618 with a telescope made by Thomas Fienus under the title *De cometa anni M.D.C. XVIII.* In this work, Froimond mentioned Kepler and Copernicus but rejected the Copernican system, a move undoubtedly due to the decision of the Congregation of the Index to prohibit *De Revolutionibus* "until corrected." Not much is known about his life between 1618 and 1626 when he published his *Meteorologicorum libri sex* (1626), a work in which he showed himself a hearty opponent of Copernicus. Later it seems that he became more interested in theology under the influence of Cornelius Jansen's attempt to revive a particular interpretation of Augustinianism. This interest climaxed in his receiving the *Doctor Theologiae* in 1628, when he also began to study Hebrew, Greek, and patristics. Even when Froimond replaced Jansen as the *Professor Ordinarius* at Louvain, he still devoted himself to the sciences, a varied training that no doubt gave him confidence in his ability to answer both the astronomical and theological claims of Lansbergen's system.[59]

In the *Anti-Aristarchus* (1631) Froimond ridiculed Lansbergen's charge regarding the physical impossibility of the diurnal motion of the fixed stars. It required greater credulity to suppose that the earth rotated every twenty-four hours because all the empirical evidence was against it. A particularly damaging argument came from artillery shots. If the earth were turning to the east, as the Copernicans supposed, then a shot from a cannon toward the east would fall far shorter than one fired to the west, assuming that shots in both directions had the same amount of gunpowder, weight of the ball, and the elevation of the cannon. But this disagreed with experience, for no matter in which direction one fires the shot, the distance is the same.[60] To buttress this argument, Froimond cited artillery experiments performed by Wilhelm IV of Hesse and Rothmann, which Tycho reported and which Kepler cited approvingly in *The Epitome of the Copernican Astronomy*. These experiments showed that an artillery shot from a cannon does not fly as quickly as Lansbergen supposed. He simply increased the speed of the ball too much.[61] Froimond exulted in his assurance that Lansbergen did not understand the physics of terrestrial motion.

Froimond also attacked the theology of Lansbergen's *Commentationes* directly, as well as targeting other Calvinists who joined the heliocentric party, especially singling out Kepler's discussion of biblical hermeneutics in his introduction to the *Astronomia Nova*. For Froimond it did not matter that Kepler was a Lutheran (to be sure with Calvinist leanings) because the essential point lay in the Protestants' championing of heterodoxy in both natural philosophy and theology. The close association in Froimond's mind shows itself in his first chapter which, instead of launching into a discussion of terrestrial mobility, sought to refute the Protestant doctrine of justification by faith alone. Perhaps more than any other anti-Copernican, Froimond united issues of natural philosophy and exegesis into a seamless garment of criticism. He saw Calvinism and Copernicanism as two sides of the same coin, the faulty methods leading to heresy in theology and causing the same in natural philosophy. Froimond, in his commentary of the General Epistles of the New Testament, often took on the Calvinists but nowhere more than in his comments on James 2:14–28 and the doctrine of justification by faith and works. Calvin's exegesis took James's words as speaking of an appearance of faith (*umbra fidei*), not true faith, when the text emphasized the fruitlessness of believing without consequent works. Froimond criticized Calvin as not taking the words of the text seriously because it emphasized that justification without works was impossible (James 2:24).[62] As we shall see, Froimond similarly charged Lansbergen with not taking seriously the plain meaning of the biblical texts that denied the motion of the earth.

What was the source of Lansbergen's errors according to Froimond? Lansbergen's accommodationist argument reveals his hermeneutical weakness. Lansbergen noted that the biblical language of movement ("God ascended") and passions ("God is angry") with reference to God could not be understood rigorously since one knows that God is immobile and impassible. Froimond no doubt also knew that this kind of language had been taken as anthropomorphism in the history of the church but Lansbergen extended this interpretative strategy to the motion of the earth, claiming that it too must be understood figuratively so that one could not exclude the earth's motion by an appeal to the literal meaning of such words. Such an extension seemed legitimate to Lansbergen; Froimond flatly rejected it. How did Froimond refute this inference? Why was this line of argument not a legitimate extension in his judgment? Froimond argued that since we know that God is literally incapable of movement and passions, we also know that such language must be taken figuratively. Such an interpretative conclusion does not hold in the case of the earth's motion because there is no independent reason to think that the earth is actually in motion. In the case of God, taking the language in a physical sense would lead to absurdity, giving God human features. In the case of terrestrial motion, the opposite is true. Not taking the language physically leads to absurdity, that of a moving earth.[63]

Froimond's refutation relied on close analysis being given to the details of each disputed text. The language of Psalm 19:6, speaking of the sun running its course, could only imply a physical sense, according to Froimond, if one were to take Scripture seriously. Froimond castigated Kepler for minimizing the power of the words when the latter emphasized the "like" (*tamquam*) of verse 6 as indicating not an absolute statement of fact but a simile for comparison. Here Froimond insisted that the text spoke of the sun running its course, not "something like running its course."[64] This attention to detail is even more evident in Froimond's treatment of Ecclesiastes 1:4, "a generation went and a generation came but the earth stands forever. The sun rises and sets and returns to its place. And there once it is reborn, it moves through midday." To dismiss this precise language as due to appearances only, as Kepler and other Copernicans customarily did, was to miss the intention of the Holy Spirit: "It is clearly unusual that by the words 'to turn, to rotate' Scripture does not indicate a true motion but only an image of motion."[65]

Froimond read the true intention of the divine author by appealing to the next verse that spoke of rivers flowing into the sea, a verse that can and must be taken as a true description of physical reality. No less, then, must descriptions of the sun's movement be so taken, unless one admits an equivocation in the words of Holy Scripture. How strange that one verse would not be speaking physically and next would be! Surely, Kepler's explanation does violence to the scriptural words. The most celebrated text in Joshua 10 demonstrated to Froimond's mind the absolute necessity of taking the plain words of the Bible in their ordinary signification and the absurdity of reading this text as only indicating appearances. If Joshua were a Copernican, he would have said, "Earth stand still!" not "Sun stand still." To attribute a Copernican meaning to Scripture would require the belief that Joshua erred in his use of language, or what is worse, that the Holy Spirit directly erred since the biblical narrator under the Spirit's inspiration described the sun and moon as standing still.[66]

In Froimond's judgment, the specific exegeses offered by Kepler and Lansbergen have common roots in a misguided hermeneutics, a method that was both bankrupt and symptomatic of their deeper heretical tendencies. Like Calvin, they imagined Scripture to be speaking of appearances, whether of faith or of terrestrial motion, and this assumption led them to dismiss the Bible's relevance to natural philosophy altogether. Earlier, we noted Lansbergen arguing that principles of natural philosophy should not be drawn from Scripture because it used inaccurate values for mathematics. Froimond turned this argument on its head by agreeing that Scripture employed "round numbers" when it spoke of pi as 30/10.[67] From this, one should not infer the Bible's irrelevance for all natural questions but only that Scripture has not attempted a precise value for the relation of the circumference of a circle to its diameter. What then can be drawn from Scripture for natural philosophy? According to Froimond, not a fullblown system of philosophy but the rudiments of such knowledge as can be found and refined by further investigation.[68] If Scripture contained something relevant on a particular natural issue, however rudimentary, this information became relevant to the discipline concerned.

Lansbergen warned against the use of Scripture in natural philosophy by indicating the mistakes committed by the church Fathers, who sometimes cited biblical texts to argue against natural truths later shown to be true. There may also have been an additional implication that because of these mistakes the church Fathers did not hold any authoritative status in natural philosophy. If Lansbergen adopted what was by the early seventeenth century a fairly standard position among Calvinists, the Fathers would not have held any authoritative status even in strictly theological matters. Froimond's Catholic commitments would surely have seen this neglect of the Fathers as a fatal mistake:

> It is manifestly false to say that nothing in Scripture refers to the knowledge of natural things when one reads in the commentaries of the Holy Fathers about the first days of the creation of the world, Ecclesiastes, the Book of Job and certain other things scattered here and there. What? Didn't Saint Ambrose, following Basil, conclude that the nature of heaven is not solid but is subtle like smoke?[69]

If Lansbergen had looked to the church Fathers rather than Calvin, he would not have been led into such errors. To Froimond, Lansbergen and other Protestants misinterpreted the Scriptures because they ignored the Fathers, their heretical bent being traceable to their audacious attempt to establish the meaning of Scripture apart from the meanings adopted by the Fathers.[70] Lansbergen, following "that Keplerian way" (*via illa Kepleriana*), committed the culpable error of demeaning the Scriptures by departing from the unanimous witness of the Fathers. Unwittingly, Lansbergen thereby attributed error to the Holy Spirit speaking in Scripture.

How could Froimond absolve Augustine or Lactantius of error? In their case, he had no need to defend the absolute inerrancy of a particular Father because the Catholics, like the Protestants, could admit many individual errors in the patristic witness. But St. Augustine's error does not justify a wholesale dismissal of biblical relevance.[71] In fact, closer inspection shows an essential difference between Augustine's and Lactantius's errors, the latter's mistake being more profound than the former's. Augustine did not deny the existence of antipodes but only that he could not imagine how they could be inhabited so quickly in human history. Augustine made an error in empirical judgment but was subject to correction.[72] Consequently, these examples only show that the Fathers sometimes erred but that those errors were not fatal to their use of Scripture in natural matters.

These examples do not justify ignoring a whole range of other patristic positions and arguments that are relevant: "the divine testimonies should be understood, not in the manner of the philosophers nor the Copernicans but 'how the Saints have understood them, thus should they be.'"[73] Froimond peruses the Fathers to discover that they have in fact delivered on natural questions so much that one must, following their example, admit the relevance of Scripture to natural philosophy. Lansbergen had argued that Moses and Daniel did not learn the foundations of the human disciplines from the Holy Spirit but from the Egyptians and Babylonians directly. He took this as evidence that the Scriptures were not intended to teach natural philosophy. Froimond's response to this argument indicates the deeper differences between them by his argument for an interpenetration of Scripture and natural philosophy. Detailed data and theories found in historic scientific documents are not found in the sacred writings, but the Holy Spirit, who inspired the sacred authors, chose to include some natural truths in passing (*obiter*) that serve to fill out the truths found in secular books. This makes the Sacred Scriptures essential for having a full-blown natural philosophy because the statements contained in them supplement natural truths. Moses and Daniel are not examples of separation but of supplementation.[74]

JACOB LANSBERGEN'S *APOLOGIA* AND FROIMOND'S REJOINDER, *VESTA*

Such a virulent attack on Copernicanism and the Reformed faith as Froimond's could not go unanswered. Two years after the appearance of Froimond's *Anti-Aristarchus* in 1631, Jacob Lansbergen, the son of the Middelburg minister Philipp, took up the battle in his father's behalf against the imposing force of the Louvain professor. Jacob, a medical doctor in Middelburg, was well trained in the relevant disciplines of astronomy and natural philosophy as well as being instructed, no doubt directly by his father, in the pertinent theological dimensions of the issues.

Jacob Lansbergen's *Apologia* (1633) was divided into six tracts, each dealing with a major aspect of the Copernican question and the third treating theological issues.[75] One can detect the underlying unease that Froimond's accusations of heresy must have produced, for in the last tract of this work Lansbergen attempts to show that the Copernican position in fact gives rise to a greater display of the glory of God and to "pious meditation." For this pious Reformed doctor, no less than for his pastor-astronomer father, "the heavens declare the glory of God" (Ps. 19:1), and the hypothesis of a mobile earth demonstrates the divine wisdom no less than the contrary thesis of a stable earth:[76]

> Because the glory of God consists in the infinite and incomprehensible divine essence and the attributes of his perfection and excellence, it is clear that this hypothesis greatly adorns the glory of God nor does it lessen the perfection of his essence and it adorns the excellence of his attributes. Nature establishes and exclaims that the Copernican system extols the wisdom, power and infinite goodness of God far more than the Ptolemaic or Tychonic hypothesis.[77]

The order of the spheres and the symmetry of the whole world-system exhibit divine wisdom while the distance of the eighth sphere gives the appearance of a nearly infinite universe, an unmistakable sign of the Creator's power.[78] Nor is the goodness of God lacking testimony, for the Copernican system demonstrates that God cares for his people, as Deuteronomy 32:11 indicates, by the movement of the earth through the ecliptic, a sure sign of God's love for humankind (*philanthropia*).[79] Pious minds grasp these natural realities and derive spiritual benefit from them. The contemplation of the celestial realm made possible by the earth's annual motion through the ecliptic should evoke in the human observer the hope of possessing heaven in the life to come.[80]

If Jacob Lansbergen claimed a doxological superiority for the Copernican system to ward off the charge of impiety, such claims to glorify God certainly did not answer Froimond's physical or even theological objections. In answering the latter objections, Jacob made full use of his father's works, citing his 1618 *Progymnasmata* and his *Uranometriae* of 1631 as well as the *Commentationes*. Philipp Lansbergen had espoused a proportionality of motion whereby the smaller inner spheres (e.g., Mercury, Venus) moved more quickly than the larger outer spheres (e.g., Saturn, eighth sphere). This, of course, was confirmed by the sidereal period of each planet's orbit. On this basis he argued that it was absurd to account for diurnal motion through the movement of the eighth sphere, as Ptolemy had done, because this would make the largest sphere move the most rapidly.[81] It was much more rational to attribute such a motion to the earth. Froimond cited the problem with artillery experiments that showed that the distances of shots were the same no matter in which direction the shot was fired (east or west). Jacob Lansbergen's only possible retort reiterated what his father had said, that the speed of the earth's motion and that of an artillery shot were the same (both covered $\frac{1}{16}$ part of a mile per second). This would explain why the direction of a shot made no difference in the distance transversed. Jacob disclaimed that his father increased the speed of a ball eightfold, as Froimond charged, because a ball shot that covered a distance of $\frac{1}{2}$ German mile would be eight times greater than one that covered only $\frac{1}{16}$ of a mile. Froimond assumed that such a ball had a uniform speed. The younger Lansbergen argued that the first $\frac{1}{16}$ of the mile would be transversed more quickly (with a tenfold impetus) and that the impetus would decrease with distance. So, Froimond's objections simply carried no weight. The overriding implausibility of the motion of the eighth sphere made the earth's motion the only possible explanation for the alternation of day and night.

Jacob Lansbergen aimed simply to show that the Copernican system "was not repugnant to the Holy Scriptures," not to prove it by them. In defense of his father's separation thesis, Jacob reasserted that the "intention of the Holy Spirit in Scripture was not to lay open the secrets of astronomy but to show the one way of salvation."[82] If Froimond wanted to take Scripture literally, he should have taken the words of the Apostle Paul seriously when he said, "all Scripture was profitable for doctrine, reproof, correction, and instruction in righteousness" (2 Tim. 3:16). Here the Apostle "expressly intends a limit" on the purpose of Scripture. It can and should be used for drawing dogmas of faith but must not be used for instruction in geometry and astronomy.[83]

How could this question of the intention of the Bible be resolved? Philipp Lansbergen and Froimond differed over the implications to be drawn from the acknowledged imprecision of language used in the Bible. According to the younger Lansbergen, Froimond had missed the obvious implication in the imprecise value of the circumference/ diameter ratio in the Bible. Such an imprecise statement not only meant that the Bible spoke in "round numbers" but that geometry, a precise science by its very nature, was not intended in the Bible.[84] The argument over imprecise biblical language connects the intention of the Holy Spirit to the issue of accommodation. Jacob Lansbergen considerably enlarged and strengthened his father's accommodation thesis with his own litany of theologians who taught implicitly that:

> God in Scripture does not deal subtly and philosophically with us but in an almost rude manner and as it were without any diligence through a certain kind of crass reason, i.e., so things can be known by a kind of common sense, both by unlearned and learned because he wanted to be a teacher not only of the educated but also of the uneducated.[85]

Citing Ambrose, Bernard, Augustine, and Aquinas, Lansbergen deflected criticisms of the Reformed religion by showing at length that Calvin (*Calvinus noster*) simply followed in the footsteps of many early Fathers when he contended that the language Moses used in Genesis was more adapted to our sight than to the actual stars.[86] Lansbergen had a very effective argument for Froimond when he could also cite the Jesuit theologian Gaspar Sanctius's endorsement of the accommodation strategy. Yet perhaps the most effective argument was Lansbergen's counterfactual one. Suppose we do attempt to derive scientific principles from Scripture, what would the conclusion be? One must conclude that the cause of thunder cannot be known because Job says so.[87] Or, even more to the point, one would conclude that the moon was larger than the other celestial bodies because Genesis 1:16 says that "God made two great luminaries." These examples demonstrate the absurdity of attempting to build geometry, or any science, on the words of Holy Scripture.

In the *Anti-Aristarchus* Froimond had confidently maintained that the Calvinists were departing from the practice of the church Fathers by their principle of irrelevance. Jacob Lansbergen, however, took great pains to show that "the wiser Fathers" always held the Scriptures to be in agreement with "true and certain philosophical dogmas" and never "repugnant to true reason." Augustine of course provided the key example by his attempt to show that the meaning of the Bible was not contrary to true proofs (*veracibus documentis*).[88] Selective treatment among the Fathers had deeper roots in the Reformation handling of the patristic tradition. Calvin, like Luther and Melanchthon, held the Fathers in high regard, but he endorsed their teachings only when he judged them to be in conformity with Scripture. Lansbergen, arguing much as Galileo did, looked back with reverence to the Fathers but found their only authority in the openness that he discovered in some, an attitude of assurance that natural truths and the Bible would never truly contradict.

Froimond was not to be outdone by the Reformed minister and his physician son. One year later the Louvain theologian responded with *Vesta* (1634), an extended attempt to answer Jacob Lansbergen's arguments and accusations point by point. This work, divided into five tracts corresponding to main sections of Lansbergen's work, exhibits Froimond's continuing perception of this debate as a theological one, since he again included in his subtitle the addendum, "in which the decree of the Sacred Congregation of the Cardinals in the years 1616 and 1633 is again defended against the Copernican movers of the earth."[89]

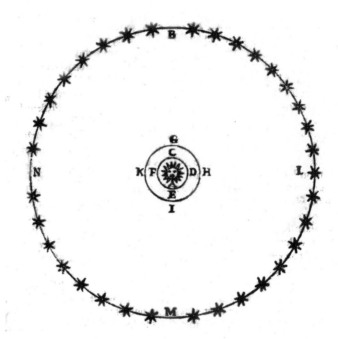

From Libert Froimond's *Vesta* (1634) showing the immense space between the sphere of Saturn (GHIK) and that of the fixed stars (BLMN) required in the Copernican system. In Froimond's mind, this useless space shows the absurdity of Copernicanism. Courtesy of the Bibliotheek van de Gemeente Universiteit te Amsterdam.

Jacob Lansbergen had read 2 Timothy 3:16 as giving an express limit to the range of subjects on which the Scriptures could be taken as an authority. How strange that a text that mentions nothing about a limit on its applicability is taken as restricted to pastoral concerns, a move that Froimond took as an arbitrary limitation. Such a restriction could even be seen as denied by the following verse (v. 17) where Paul admonished Timothy to be a man of God, a task that would inevitably involve knowledge of human concerns. Froimond's view that Scripture and natural philosophy supplement one another was restated as a theory of seed principles:

We think that it is not the principal purpose of the Holy Spirit to disseminate the human disciplines by the text of Sacred Scripture.

But there are certain kinds of seeds of these [disciplines], mixed with divine things, that slip out at one time or another. The Holy Spirit is everywhere holy and true, nor can he let any lie ever slip out whether in little things or big ones.[90]

Froimond's language indicates the severity of the departure he perceived in the Lansbergens' argument, nothing less than attributing prevarication to the Holy Spirit.[91] Nor was Lansbergen consistent with his espoused principles because he himself quoted texts of the Bible dealing with the natural world when he had maintained that natural truths could not be drawn from it.[92]

Froimond recognized the difficulty in refuting the Lansbergens' accommodationist argument, for no matter what the text of Scripture said, they could always assert its irrelevance by claiming that it spoke in everyday language and was not teaching natural truths.[93] But Froimond saw this gambit as ignoring the obvious differences of language between poetic comparisons and straightforward narratives. The language of Ecclesiastes 1 contained both. It spoke of the sun "like a bridegroom proceeding from his chamber," and the comparison demanded a nonliteral understanding. It also narrated the movement of the sun "with a simple and fitting language," and the author used the same kind of speech for the rivers that flowed. Here was no indication of poetic comparison, no reason to take the text nonphysically.[94] The debate over accommodated language appeared to be at an impasse, but Froimond was sure that the Copernicans' position on biblical language was motivated by a prior commitment to heliocentrism that wreaked havoc on their approach to the Bible. Not surprisingly, the Copernicans' prior commitment to Calvinism had done the same thing. Those commitments also distorted their vision of the church Fathers who were not to be treated simply as human authorities. When they were unanimous in their endorsements or condemnations, they should be followed without question. Supposing that we know who the "wiser Fathers" are is presumptuous.

The protracted exchange between the Lansbergens and Froimond was colored much by the dynamics of Catholic-Reformed polemics. Whether this reflects the debate in the Netherlands as a whole in the 1630s is unclear because of the unavailability of documents. But the use of good science to bolster true religion was shared by both sides of the debate. From Froimond's earlier cosmological work we would be justified in concluding that he might have employed Copernicanism as an ally, but he was above all loyal to the church and he strenuously defended its decisions after the 1616 declaration in Rome. By their own principles, the Lansbergens had to be convinced of heliocentrism on astronomical and physical grounds, but it is easy to misconstrue the nature of their beliefs. Both argued that exegesis was irrelevant to the question of the earth's motion, but they both also believed that the universe was a declaration of the glory of God. The theological truths proclaimed in Scripture were embodied also in nature. The underlying Roman Catholic–Reformed polemics seemed to control the debates prior to the appearance of Cartesianism, but the available literature after 1640 shows one thing clearly: Reformed thinkers now had a new reason to be divided among themselves.

THE CONJUNCTION OF COPERNICANISM AND CARTESIANISM

With the introduction of Descartes' mechanical philosophy into Dutch intellectual life (ca. 1640) the debates over Copernicanism took on a new dimension, one that would join together the fates of both scientific developments in the minds of advocates and opponents. The Cartesian philosophy became the handmaiden to theology for some adherents of the Reformed religion while it was bitterly opposed by others confessing the same religious principles.[95] As the new faith of the Reformation made its way into Dutch life, many unanswered questions arose over how theology should relate to philosophy and even over what authority theology played in determining the faith of Christians. In two of the major cities (Leiden, Utrecht) where universities had been founded in the early modern period these debates were carried on by clergy and other professional leaders who hoped to establish the still young Dutch Republic on a solid educational footing.

The faculties of philosophy and theology at Leiden eventually became a haven for Cartesians, but not without considerable struggle. Founded in 1575, by the turn of the seventeenth century the university had already become known as a place where wide diversity of viewpoints was tolerated, the differences between Jacob Arminius and Francis Gomarus, the anti-Remonstrant, being a case in point.[96] The burgermeisters of the city never allowed the strict Calvinists ("the precisians") to dominate even the faculty of divinity, much less the arts, medicine, or law faculties. Indeed, they resisted any control of the university by the church, although Leiden was not immune from the effects of the crisis of 1619 occasioned by the Synod of Dordt.[97] Even with the growing influence of the precisians, however, Leiden remained a university of toleration and sought to balance its faculties with opposing viewpoints. Descartes' ideas had been adopted in varying degrees by the philosophers who saw his work not so much as an innovative dogma to replace Aristotelianism but who saw Descartes himself as an advocate of independent research.[98]

That same independence of thought characterized the theology faculty in which Johannes Cocceius was able to develop his historicizing approach to the Bible, an approach that was to affect his conception of the Sabbath and that brought him into fierce conflict with Gisbert Voet of Utrecht.[99] Cocceius himself, who came to the Leiden faculty in 1650, did not seem to be all that knowledgeable about Cartesian thought, but his disciples were deeply influenced by it.[100] Perhaps this was due to his traditional view of philosophy as a handmaiden even though he did not consider Aristotelianism as the right choice for that role. Cocceius wanted to separate the domains of philosophy and theology, the former depending on proofs from reason while the latter depended on proper exegesis of the word of God:

> Is there much of profit in philosophy, whether Platonic, Aristotelian or eclectic or of some other type of philosophy? We do not despise philosophy at all. Rather we praise it highly in Christians and even admire it in non-Christians. But we say this. Although those gifts are of great value in faithful men, yet if any philosopher who is full of his own wisdom approaches Sacred Scripture and brings some hypothesis to it wanting to confirm it from Scripture, he will be nothing other than a distorter of God's words. . . . He rather should expect to be taught from the Scriptures as a child.[101]

Whatever separation Cocceius may have advocated, Cartesianism and heretical theology were bound together in the minds of Cocceius's opponents. This required the leading exponent of Cartesianism at Leiden, Abraham Heidanus, to defend his colleague's distinction between the goals and methods of theology and philosophy. Heidanus had to expend considerable effort to disabuse the public of the intimacy of Cartesianism and Cocceius's theology:

> For a long time the world has wanted to make us believe that the theology of Dr. Cocceius and the philosophy of M. Descartes was so closely tied to one another that each entire system was modeled on the other and that all the scabs and venom of one have rubbed off on the other. As none can pass for a Cocceian unless he sides with the Cartesians, no philosophy is more fitted to serve as a handmaid for theology than the Cartesian.[102]

Heidanus found the handmaiden metaphor useful for his purposes, with an emphasis on a mutually supportive role for each system. "A good union and relationship" should be sought, not a submission of one to the other. Natural reason will be sanctified by revealed truth as each remains properly within its own domain and at the same time looks to the Father of Lights for illumination.[103]

Cartesianism also found a warm reception in the late 1630s at the newly established University of Utrecht, especially in the medical faculty, but there it also had a more tenacious opponent than any in Leiden, the leading Reformed theologian Gisbert Voet. Having taken his *doctor theologiae* under Gomarus at Leiden, Voet became head of the Illustre school that was founded in 1636 and was almost constantly involved in controversy over philosophical and theological matters. Three of the most prominent disputes included his open debates with Descartes and the Cartesians, those with the Leiden theologian Johannes Cocceius, and his apologetic attempts directed against the leading Roman Catholic of Utrecht, Libert Froimond.[104]

In his inaugural address (*De pietate cum scientia conjugenda*), Voet displayed his conviction of how knowledge and piety must cooperate in the advancement of the gospel. He argued vigorously for the necessity of philosophy to undergird the Scriptures and the true Reformed faith. In 1641 Voet became the rector-magnificus, and all faculty members had to submit their proposed lectures for his approval. The leading medical professor, Henricus Regius, presented to Voet an outline of physiological lectures to ask if any damage to theology would result from teaching the Cartesian philosophy. During the remainder of that year, the Cartesian system was debated throughout the university, and by the end of the year (8 December) Regius presented the thesis that the union of body and soul is not an essential but accidental connection.[105] Voet responded by denying the philosophical validity and the theological acceptability of Regius's theses, also including arguments against the motion of the earth. Voet argued that philosophical inquiry should undergird the truth of Scripture but that ultimately Scripture must judge any philosophical system. Not only were faith and morals taught in Scripture, but the foundations of all philosophy were implicitly there as well, a belief in the "Mosaic and sacred physics" with which the non-Aristotelian physics of his day could not be reconciled. Because Descartes' system of nature denied substantial forms, the very foundations of Christian doctrines (e.g., Trinity, Incarnation) were threatened. This intimacy between Aristotelian natural philosophy and theology was urged earlier in Voet's *Thersites Heautontimorumenos* (1635) when he argued that Psalm 104, for example, contained sections not only on theology but on history, politics, economics, geography, physics, astronomy, chronology, mechanics, and navigation.[106] The accommodation argument adduced by Lansbergen and furthered by the Cartesians defined the relevance of Scripture with respect to those things which pertain to religion, not to natural philosophy (*quae ad religionem non pertinent*). Voet found this argument unconvincing because it was not clear when something in Scripture pertained to religion and when it did not. Further, if the Scriptures spoke in language that was not accurate enough for philosophical discourse, then in Voet's judgment this simply meant that Scripture was false, deceptive, and unsuitable. Unlearned readers of the Bible too, encouraged as they were by the Protestants, would be led astray by the inaccurate words of Scripture.[107] Such distinctions wreaked havoc on the true Reformed faith by undermining the ultimate authority of Scripture.

Around 1650 Utrecht and Leiden again became the centers of exchange in the issue of terrestrial motion, with a vigorous debate between the Leiden preacher Jacob de Bois and the Utrecht physician Lambert Velthuysen. Little is known of de Bois's life, but he did come to Leiden as a minister in 1646 and remained there until his death in 1661. In the late 1640s he published two defenses of infant baptism against the Anabaptists of the city before entering the controversy over the Cartesian philosophy.[108] Velthuysen lived most of his life in Utrecht, where he was trained in philosophy, theology, and medicine, later becoming one of the most outspoken advocates of the Cartesian philosophy in that city. In 1655 de Bois attacked the Cartesians with his *Naaktheid van de Cartesiaansche Philosophie* ("The Barrenness of the Cartesian Philosophy"), loudly proclaiming his conviction that both the movement of the earth and the Cartesian philosophy as a whole were contrary to Holy Scripture. De Bois specifically targeted one of the leading Cartesian theologians, Christoph Wittich, who had written a short work on the style of Scripture and argued that its characteristic mode of expression did not permit philosophical conclusions to be drawn from it.[109] We shall return to Wittich in due course. Velthuysen responded with a work in Dutch entitled *Proof* (*Bewys*, Latin title *Demonstratio*), which defended the motion of the earth and Cartesian philosophy from the theological accusation that these stood in contradiction to Scripture. De Bois's opposition only grew stronger with *Schadelijkheid der Cartesiaansche Philosophie* ("The harmfulness of the Cartesian Philosophy"), and Velthuysen answered de Bois's criticism with a *Nader Bewys* ("Further Proof").[110]

The tone of the Velthuysen-de Bois interchange resulted from the distance that existed in the 1650s between the Reformed Church in the Netherlands and the Roman Catholic Church. No longer was Velthuysen worried, as the Lansbergens had been, about how to answer the Catholics in these matters but simply about how to resolve the question of the motion of the earth within the Reformed Church.[111] The acrimony had become intense within the Reformed Confessional Church, perhaps because a distinctly Protestant method of argument had now emerged, an appeal by the opponents of terrestrial motion to the "clear texts of Scripture."[112] An appeal to indubitable texts augured a devastating end to the Cartesian philosophy and free inquiry. Velthuysen's self-imposed task defended both, but he did so by showing the misuse of Sacred Scripture in natural debates.

The purpose of the *Demonstratio* rang out clearly from its opening sentence: "I do not intend to assert the motion of the earth from the sacred pages for I do not think that it can be proved from the Sacred Scriptures." Velthuysen simply wanted to prove that the meaning of those texts that had been adduced as "clear proof" of an irreconcilable conflict between the Bible and the motion of the earth had been improperly understood. To accomplish this task, he was required to consider hermeneutics explicitly and to formulate a clear thesis:

> We think that a divine and therefore equal authority can be attributed to each and every book of Sacred Scripture and their parts according to the canon received by the Reformers. But we must take not the letters or the words but their meaning as Scripture lest the sacred texts appear to contradict themselves. If we go by the words themselves without attending to the meaning—to affirm that God is a spirit and yet has hands and feet—then it involves a contradiction. But Scripture teaches both if you attend to the words only. Yet since the Reformers themselves think that no teacher is immune from errors in explicating the sacred texts, we should be warned lest we heedlessly reproach this view as impious and heretical by condemning our neighbors. We do not want to be like the Romanists who command the minds of men and arrogate power to themselves.[113]

Velthuysen argued that in explicating the meaning of the Bible, one must distinguish between the words of the text and the actual teaching or doctrine intended *by* the text, not every word in a biblical text being given equal weight for formulating doctrine and authoritative interpretations. De Bois rejected this distinction between the words of a text and its underlying doctrine, noting that it would be impossible to ever arrive at any certainty about doctrine.[114] Yet de Bois unwittingly played into the hand of Velthuysen when he sought to reconcile apparent contradictions in the Bible. The apostle Stephen in his speech in Acts 7:4 says that Abraham left Haran and entered Canaan after the death of his father, but Genesis 11:31 and 12:4 indicate that the patriarch left Haran before his father died. De Bois took the Genesis texts at face value and attempted an explanation of the Acts text by offering two possibilities, one hermeneutical and one textual, favoring the explanation that Stephen had spoken according to the common Jewish beliefs held by his audience. De Bois went so far as to say that Stephen, being led by the Spirit of God, "did not err in his teaching and said nothing repugnant to truth. He [Stephen] either did not have a perfect knowledge of the historical circumstances or he did not attend to them."[115] This contradiction was not lost on Velthuysen, who drove home his point with a vengeance—"so the Holy Spirit taught a pure lie through Stephen's mouth and affirmed what is abhorrent to truth."[116] To Velthuysen, de Bois must either hold to his insistence on every word having equal weight of authority or admit, as his example of Stephen required, that a distinction must be made between the words of the text and the teaching of the text. Velthuysen would have none of this pontificating, for de Bois was simply constructing dogmas that carried no weight for Christian interpreters. The authority of Scripture should not be treated in the manner of the Roman Catholics, with their arbitrary dogmas.[117]

In Velthuysen's approach, the resolution of apparent contradictions in the Bible should be resolved by using reason, but the content of that reason was not to be the visible words of Scripture but the meaning of the text. In good Cartesian fashion, the underlying meaning of a text could be perceived clearly and distinctly (*clare et distincte*), a sure path to truth, although one must weigh carefully the actual words.[118] Velthuysen then explains the motion of the earth by this method. If the words of Scripture were clearly and distinctly perceived on the matter of solar motion and terrestrial immobility, there would be no further dispute. Yet the Holy Spirit, in inspiring the sacred authors, often used language whose proper meaning could not be directly inferred from the words themselves. In that case, the divine author did not wish to offer dogma to teach, affirm, or deny.[119] The circumstances around the texts (times, locations, customs) provide the context by which to decide when the words of the text should be taken as teaching dogma. Consequently, texts like Psalm 19 and Joshua 10 do not offer directly any dogmas about solar motion but are conditioned by customary modes of speech found in the Bible.

The separation of philosophy and theology was furthered in Velthuysen's second work *Luculentior Probatio*, where he delineated the former as a handmaiden (*ancilla*) of the latter, the one depending strictly on reason and the other on the word of God as the rule of faith (*regula fidei*).[120] This time Velthuysen limited theology even more, insisting that only the word of God could be taken as authoritative and not any human exposition of it. The ultimate decision as to the meaning of a text of Scripture rests with the knowledge and experience of the individual Christian.[121] Velthuysen redirected the Reformed belief in the fallibility of church councils—originally used by the magisterial Reformers against Roman Catholic opponents—to limit the decisions of the Reformed churches themselves. Such synods and classes had no authority to add conditions of faith to what was already a well-defined Reformed faith. Consequently, neither de Bois nor any other Christian should attempt to make the question of the earth's motion into a dogma of faith or to pronounce one or the other side heretical. This was a matter of individual faith and decision. Velthuysen's argument, of course, expanded far beyond the issue of terrestrial motion into a debate over the freedom of philosophizing and the use of reason in theology. His was an argument that had moved far beyond the original Reformed claims into a rigid separation of philosophy from theology that severely limited the role of the Bible in cosmology, an argument that would be even more widely expanded in the magnum opus of Christoph Wittich.

With the advent of Cartesianism in the Low Countries, the debates over Copernicanism took on wider dimensions that intensified problems about the relation of natural philosophy to theology. The motion of the earth was embedded in a larger framework of discussion shaped by the problematics inherent in the Reformed view of the Bible. Since the Bible was the sole and final authority of faith for Reformed thinkers, it raised the problem of how far that authority should extend. Those of biblicistic tendencies like Voet and de Bois not only saw an implicit cosmology in the Bible, they extended scriptural authority over every possible question that could be posed. On the other hand, Cocceius and Velthuysen, attending to the everyday language of the Bible, tended to separate its intention from that of philosophy. By doing so, they limited the scope of scriptural authority more severely, a move that was made possible by their rejection of the infallibility of church councils. Velthuysen particularly, as we have seen, argued for an unlimited freedom of the individual not only in philosophy but in the interpretation of Scripture itself.

CHRISTOPH WITTICH'S DEFENSE OF COPERNICANISM AND CARTESIANISM

The arguments for the separation of natural philosophy and theology were most thoroughly developed during the 1650s by the expatriate Silesian theologian Christoph Wittich. Wittich was born in 1625 in Breig, Lower Silesia, where his father was the superintendent of the Protestant church in the dukedom. In 1642 he began his university studies in law at Bremen but soon shifted his focus to philosophy by studying and adopting the Cartesian philosophy. Four years later, he matriculated at Groningen where he also became an adherent to what was by now traditional Reformed theology, although his main studies were still natural philosophy. In 1646 he defended a physical disputation entitled *On Water* (*de Aqua*) under Martin Schoock and then moved on to Leiden to further his studies. In 1651 he became a professor of mathematics at Herborn while also teaching theology. Beginning with these years, he began to consider the relation of natural philosophy to the Scriptures, writing two disputations on the misuse of Holy Scripture in philosophical matters as well as endorsing the movement of the earth.[122] Then in 1655 Wittich became professor of theology at Nijmegen, where he taught a wide variety of subjects, and where he was obliged again to address the problem of natural philosophy and the use of reason in understanding the Scriptures. When he moved to Leiden University in 1671 as professor of theology, he delivered his inaugural address under the title *Theologia Pacifica* because the crosscurrents of theological debate and opprobrium had filled the Dutch air for a long time. Wittich's approach to the problem of Scripture and natural philosophy consistently sought to frame any discussion in a spirit of mutual good for both disciplines.[123]

Wittich's irenicism followed the argumentation set out by the Lansbergens in which the separate and different methods of philosophy and theology were given their legitimate spheres of operation. Now, however, his Cartesianism allowed him to ground his view of the disciplines in an explicitly epistemological framework. By the time Wittich wrote his magnum opus, *Consensus Veritatis*, in 1659, he had constructed a rigid distinction between common, everyday knowledge and philosophical inquiry:

> I use an explicit difference between a so-called "vulgar" knowledge that is common to all men and philosophical knowledge often called "accurate" to demonstrate by argument that the purpose [*principium*] of Sacred Scripture is not this philosophical knowledge.[124]

The distinction between two different types of knowledge was grounded in two separate epistemic principles. Philosophical or accurate knowledge was based on "the light of reason" which, if properly employed, could lead to the discovery of many "obscure truths."[125] Such truths could not be discerned by the senses, only by clear and distinct ideas. On the other hand, there was that vulgar knowledge that was common to all humans and that did not require any analysis of experience or long chain of reasoning to attain. Here the light of reason operated, but only in a nonanalytic fashion. Scripture was addressed to all in the language of sense experience since the Holy Spirit had no intention of making everyone a philosopher. Philosophy and theology therefore had completely separate tasks and spheres of operation without any overlap between their methods or results. Wittich the theologian answered the charge that Descartes' method of systematic doubt could be used to undermine truths of theology by insisting that this doubt extended "to neither matters of faith nor manners of life."[126] The Scriptures then had no relevance to any questions posed by the natural philosopher, including the motion of the earth, because the holy writings were limited to sense experience while true philosophical knowledge rested on reasoned analysis.

A reasoned analysis of the earth's motion was needed. Since the Cartesian system demanded necessary demonstrations modeled on Euclidean methods, Wittich turned to physical proofs that went well beyond the standard astronomical evidence used by earlier Copernicans. But the spirit of Cartesianism also demanded that he derive the physical principles of terrestrial motion from a deeper metaphysics, one that entailed a complete cosmogony. Following Descartes, Wittich endorsed a plenum universe in which there could be no vacuum, no empty space without even the smallest corpuscles present. The heavens were filled with a pervasive celestial matter (*materia coelestis ambiens*) that operated on the celestial bodies to move them in their respective paths. The Cartesian theory of motion required some such corpuscles because it did not allow for action at a distance as Newton's theory of gravitation was later to do. Something physical had to do the work of moving objects. Further, physical objects could only be derived from other physical objects, so the Cartesians explained the existence of the earth by claiming that it was a former star that was spun off into its current orbit by the swirling vortices.[127]

Explaining the twofold motion of the earth incorporated the idea of surrounding celestial material, matter that comprised not only the environment but the cause of earth's motion. Wittich explained the annual motion of the earth as resulting from the vortex of the earth being absorbed by the solar vortex and then thrust out into its present position, where it was carried in a fluid celestial matter that moved in a circle. A hard body like the earth then was inevitably swept along in this swirling motion.[128] The diurnal motion of the earth that explains the alternation of day and night invoked two different concepts. The first was a residual motion from the earth being formerly a star; the second was an inner force generated by the motion of subtle matter (*materia subtilis*) in the earth's center.[129] The earth and moon share in the same sphere and motion as well as having an equal motive force (*vis agitationis*); the difference in their speed of motion has to do with the different distances they must traverse.[130] Both types of motion were conceived in very physical terms and both hearkened back to an earlier state of the universe from which their current motions derived.

In terms of theology, Wittich's huge volume focused on the issue of accommodation in Scripture, but his level of sophistication far exceeded anything Kepler, Galileo, or Foscarini had produced. His discussion was framed in openly Reformed terms since the name of John Calvin appears frequently among his arguments. Within Reformed circles, Calvin had become the new "Church Father" to whom various disputants appealed as a precedent, but Calvin's position on natural truths was itself subject to different interpretations. The discussion of

Calvin at the beginning of this chapter may have seemed straightforward enough, but since Calvin was now an authority with whom it was difficult to disagree, arguments on both sides of the issue had to appeal to him. On the surface, Wittich's position found unambiguous support in Calvin's approach to the Bible because both recognized that Scripture employed customary formulas of the ancient world that were not intended to give astronomical accuracy.[131] What is less clear in Calvin is Wittich's extension of this idea into a full-blown separation of natural philosophy and theology in which Scripture had no philosophical relevance because its use of customary formulas involved cultural prejudices. The question was posed in terms of the meaning of a phrase in Isaiah 8:1 that was normally translated into Latin as *ut scriberet stylo humano* "to write in a human style"—a phrase that Wittich took as an explicit scriptural approbation of accommodation. Attempting to give philosophical accuracy would have required the Holy Spirit to use a new language (*nova lingua*) which would have been incomprehensible to its recipients.[132] Calvin's treatment of Genesis 1:16 seemed only to confirm Wittich's position because Calvin recognized Scripture as directed to the unlearned and unskilled (*indoctis et rudibus*). However, opponents of terrestial mobility such as de Bois read Isaiah and Calvin differently. In de Bois's interpretation of Isaiah's words, to write in a human style was to use distinct, clear, and big letters that could be read by all who passed by so that the clarity and legibility of letters may be considered, not the perspicuity of meaning. De Bois invoked Calvin's authority as well as that of other Genevans to argue that the Cartesians had missed Isaiah's meaning.[133] For Wittich, de Bois was straining at gnats, for Calvin recognized that Isaiah intended both the clarity of the letters and the perspicuity of meaning. The former served the latter. In general, Calvin argued for a clarity of meaning and doctrine, but he also recognized that Scripture had limited its scope to the senses. De Bois had distorted Calvin's doctrine of Scripture.

The question then remains who had interpreted Calvin properly, de Bois or Wittich. Cartesians such as Wittich could clearly appeal to Calvin's recognition of nontechnical language in Scripture, but whether this could be expanded into a bifurcated view of human knowledge is doubtful. Calvin's appeal to accommodation was limited to the specific issues facing him so it is highly speculative to claim that he would have endorsed Wittich's rigid separation. Calvin probably thought that Scripture had considerable relevance to philosophy, especially on moral issues, and his references to a *christiana philosophia* rested on the conviction that Scripture could give answers to classical philosophical issues. On the other hand, Calvin's entire theology could be characterized as accommodationist because of his high view of God. Any divine attempt to communicate with humans entailed for him some adjustment to human limitations; the incarnation of the second person of the Trinity showed such a need clearly. So Wittich's appeal to Calvin's accommodation does reflect the Genevan father's deeper thought, although it is unclear how Calvin might have responded to Cartesianism. As is so often the case with appeals to historical precedents, the ambiguities in Calvin's works offered more than one way of looking at a new issue, one on which the original figure made no judgments. But the argument over Calvin does indicate how deeply the Reformation divide had shaped the discussions of natural philosophy in the seventeenth century. The older animosities between Rome and Geneva had faded, and the ambiguities of a new tradition were front and center.

I suggest that the theological reception of Copernican astronomy and cosmology can be explained by two factors: the structure and spirit of Dutch culture and the inner dynamics of Reformed theology. It is essential to emphasize of course that all Dutch-speaking students of Copernicus argued for his system on astronomical grounds, not theological ones. From Frisius to Beeckman, from Lansbergen to Wittich, it was a perceived coherence of Copernicus's system that won the day, and their reception of that system was often fenced off from theological criticisms by arguments for disciplinary separation. Yet the lack of theological opprobrium in the Netherlands reflects the structure of church-state relations in the northern provinces after the Union of Utrecht (1579). The seven northern provinces refused to endorse any transprovincial church, preferring to leave that question to the individual towns. Consequently, there was no ecclesiastical mechanism to censure deviant developments in either natural philosophy or theology. But that lack of national church structure also reflects the spirit of the northern Netherlands in the seventeenth century. There was always a tendency to keep partisan positions from dominating educational institutions. Though William of Orange founded Leiden University (1575) for the training of Reformed ministers, the city fathers of Leiden steadfastly refused to let one particular Calvinist group dominate the theology faculty. And even though the academy in Utrecht was dominated by the imposing figure of Gisbert Voet, the city itself allowed for free movement and publication of divergent viewpoints, both from other Reformed thinkers such as Velthuysen and even Roman Catholic bishops like Froimond.

The inner dynamics of Reformed thinking played a crucial role in the reception of Copernicanism as well. The precedent set by Calvin in his appeal to accommodation prevailed throughout the period we have covered, but there is a deeper sense in which it helped shape the discussions. Reformed thought emphasized reason in its theology because Reformed thinkers, from Calvin to Wittich, tended to see the Roman Catholic Church as riddled with superstition. This included a tendency toward deferential obeisance to authority even when that authority had no divine sanction. The Reformers rejected the teaching authority of the Roman Catholic Church in its magisterium, so they were forced to substitute some means of adjudicating interpretative disputes. Calvin had emphasized continuity with the Fathers, but he did so only in a selective fashion, a tendency that became more pronounced among the Dutch Reformed as they also appealed to reason in their theological disputes. Reasonable use of the word of God in Scripture would be the antidote to "Popish superstitions" and spurious appeals to authority. The Cartesian rhetoric of relying on clear and distinct ideas and on the primacy of reason must have appealed to many of Reformed persuasion. Indeed, as we have seen, Wittich saw a felicitous confluence of epistemology. Understanding the word of God demanded piercing beneath the appearances of language to the true meaning that was always clear and distinct. Philosophy demanded the same of the natural world. In the final analysis, I suggest that all three modern movements—Reformed theology, Copernicanism, and Cartesianism—found a home in the Netherlands because the Dutch had a spirit of openness that was engendered by their rebellion against Spanish rule. In their search for a national identity, they would not to be ruled by the Latins to the south, either politically or ecclesiastically.

CHAPTER SIX

Copernicanism and the Bible in Catholic Europe

*W*e have now surveyed the terrain in which the cosmologies occasioned by Copernicus interacted with biblical interpretation in the two major movements of the Protestant Reformation (Lutheranism, Reformed Calvinism) in a period of just over one hundred years. This material remained largely unanalyzed compared to the much more thoroughly understood situation in Catholic Europe, especially the Galileo affair and its Italian context. This scholarly asymmetry resulted both from the choices of individual scholars and from the lack of attention given to the theological contexts of Lutheran and Reformed theologies. Both forms of Reformation theology far exceeded cosmological issues, but I have explicated those aspects that were most relevant to the exegetical issues raised by the emerging cosmologies of the period.

Now we are in a position to compare exegetical methods and ecclesiastical contexts on both sides of the major religious divide that shaped early modern Europe. We have already noted how Protestant-Catholic polemics lay behind discussions in the Netherlands in the 1630s and how Lutheran and Reformed theologies shaped discussions in those areas where each was the reigning orthodoxy. How do these discussions compare with those that took place in Catholic Europe? At the level of the individual, we find a diversity of responses among Catholic thinkers, with both non-Copernicans and Copernicans arguing in ways similar to those of their Protestant counterparts. A much greater difference emerges at the institutional level.

The most famous defense of the Copernican theory in Catholic Italy came from the pen of Galileo Galilei in 1615 in his *Letter to the Grand Duchess Christina*. This eventually led to the proscription of heliocentrism, 5 March 1616, by the Congregation of the Index. The Congregation's decree explicitly prohibited *De Revolutionibus* "until corrected," Diego de Zuniga's *On Job*, and a letter published by a certain Carmelite Father under the title of *Letter of the Reverend Father Paolo Foscarini on the Pythagorean and Copernican Opinion of the Earth's Motion and Sun's Rest and on the New Pythagorean World System*. Only this work of Foscarini's was "completely prohibited and condemned" because of its blatant theological defense of the earth's mobility."[1] The letter is dated 6 January 1615, and it is likely that Galileo consulted it when he wrote his own *Letter to Christina* in June of 1615 since several of the same arguments show up in both documents.[2]

THE INSTITUTIONAL STRUCTURE OF CATHOLIC HERMENEUTICS

In Catholic Europe one finds a more complex situation than in the Protestant countries for several reasons, not the least of which is the Catholic belief in the infallibility of general councils. Since Christian antiquity the Catholic church subscribed to the notion that decisions made by the college of bishops sitting in an ecumenical council were binding on the faithful and especially on theologians who taught the dogmas of the church. This belief was summarized by the famous dictum of the fifth-century monk, Vincent of Lérins, who said that an interpretation of Scripture must be held if it is *quod ubique, quod semper et quod ab omnibus creditum est* (believed everywhere, always, and by all).[3] The three principles of universality, antiquity, and consensus were still guiding rules for interpreters of Scripture in the Counter-Reformation, although they were rooted in the faith of the Catholic church long before Vincent of Lérins expressed them in writing.

The Council of Trent reaffirmed these interpretative principles in response to the Protestant Reformation that emphasized the Scriptures alone as the final authority. The Protestant Reformers certainly considered the church Fathers venerable and worthy of their assent, but, as

The Representation of the Fathers assembled in the Council of Trent: began about the end of the year 1545. Concluded towards the end of 1563 under y Pontificate of Paul III. Iulius III. Marcel II. Paul IV. and Pius IV. There were XXV. Sessions, in which were present VII Cardinals, V. whereof were the Popes Legates, XVI. Ambassadours from Kings, Princes & Repub=licks, CCL.Patriarchs, Archbishops, Bishops, — Abbots and Generals of Orders, All Divines and Doctours of the Civil and Canon Law.

The Council of Trent in session. From Pierre Jurieu's *The History of the Council of Trent.* Reproduced from the original held by the Notre Dame Library, Department of Special Collections.

successive generations came along, the patristic tradition fell more and more by the wayside in Protestant thinking. Patristic unanimity became the controlling hermeneutical environment in Catholic Europe during the late sixteenth and seventeenth centuries. The fourth session of Trent (8 April 1546) promulgated authoritative guidelines for biblical interpretation as follows:

Further [the council] decrees that to curb the petulant ingenuity of some, no one by relying on his own wisdom should distort Sacred Scripture into his own meanings and dare to interpret Sacred Scripture contrary to the sense which the holy Mother Church has held and holds in matters of faith and morals which pertain to the edification of Christian doctrine. It belongs to her [the church] judge the true sense and interpretation of Scripture. Nor is it even permitted to hold interpretations contrary to the unanimous agreement of the Fathers, even though such interpretations should never at any time be published.[4]

The background of the Protestant Reformation is clearly discernible in this proscription of individual interpretation. This decree set the matter of hermeneutics within the context of patristic unanimity and the right of the church's magisterium to adjudicate issues of interpretation. Further, the post-Tridentine emphasis on literal interpretation to counter the Protestant interpretations of the Eucharist, penance, and the papacy set a climate in which any perceived departure from the patristic heritage was thoroughly suspect. The problems of interpreting the nature of astronomical and physical theories combined with these hermeneutical rules served to create complex context in which the issue of whether natural philosophy was a *de fide* matter had to be resolved at the institutional rather than at the individual level.

Although many exegetical similarities between Catholic and Protestant philosophers and theologians existed, the Catholic response to Copernicus differed significantly because the theological institutions of Catholic Europe were different, institutional strictures that eventually filtered down to the individual thinker. The forces at work in the Catholic church were even more centripetal than those in Lutheranism, forces that often called for unified judgment on theological matters. The Protestants could never marshal a unified response to Rome, never an ecumenical council on the order of Trent. Even the Lutheran Formula of Concord and the Reformed Synod of Dordt never had the binding force of Trent's canons. The Protestants had nothing corresponding to the Holy Office of the Inquisition, and most Protestant lands did not have the power to enforce judgments of heresy as the Catholics had, though they shared the belief that heresy was a capital crime because of its harmful effects on society. Catholicism also had a greater division of labor among theologians, philosophers, and astronomers than Protestantism had. The Catholic church leaned much more heavily on theological expertise that was centered in the Roman

Curia, which had both the institutional mechanisms and the will to adjudicate disputed questions. Thus, the process that led to the 1616 condemnation of heliocentrism was not unique or new because many theological issues had gone down the same path. The uniqueness of the process lay in its being applied to an issue of natural philosophy.

This institutional structure had its effect on Catholics writing on the Copernican issue because they shared the common belief in the necessity of submitting to the proper authorities for final adjudication. Those experts included astronomers and natural philosophers, but theologians were also called in once it was thought that the motion of the earth was relevant to scriptural interpretation. The effect on the greatest Jesuit astronomer of the late sixteenth century, Christopher Clavius, became evident in his cautious treatment of Copernicus. Clavius was hesitant to condemn Copernicus on scriptural grounds, saying only that Scripture *appeared* to contradict the motion of the earth. He quoted the standard texts (Ps. 104:5; 19:6,7) but he did not make any harsh judgments of heresy because he knew that theological judgments were in the hands of the church and its expert theologians. Clavius was prepared to say that the motion of the earth was *falsa in philosophia* because he possessed sufficient knowledge of astronomy and physics. And he probably believed that Scripture was clearly against heliocentrism, but he also knew that the judgment was not his.[5] The same tone underlay that famous letter of the Copernican Foscarini whose entire argument for the acceptability of geokineticism was submitted to "the feet of the Highest Pastor."[6] Galileo of course made similar affirmations in his *Letter to Christina*. We observed in chapter 5 how Froimond read Copernicus as an astronomical rather than a cosmological theory. He explicitly defended the 1616 decision in Rome as reflecting the best astronomy, physics, and biblical interpretation. Even more so than the Italians, Froimond sought to ground the church's decisions in reason and a proper interpretation of Scripture, a necessity borne of the polemics of battling with the Reformed heretics. All Catholics willing to enter into the debate knew that the final judgment of theological acceptability was properly in the hands of the church. This submissive tone does not appear in the Protestant writers, neither Lutheran nor Reformed. From Melanchthon to Haffenreffer, the Lutheran theologians who opposed geokineticism had only Scripture as their last court of appeal. Kepler the Lutheran, on the other hand, had to invoke freedom of judgment in interpreting Scripture. This tendency showed itself even more strongly in the Reformed Netherlands with both Velthuysen and Wittich invoking an individual right of interpretation. The lack of a unified ecclesiastical authority in the Netherlands also hindered any resolution beyond the ability to quote Scripture against one's opponent. These very different institutional contexts had profound effects on the theological dimensions of cosmological discussions. The decision of the Congregation of the Index in Rome in 1616, I suggest, would probably never have happened in Protestant lands. The Lutherans and Reformed had no formal mechanisms for such a unified response, nor were their theologies likely to produce one.

Galileo Galilei on the Interpretation of Scripture

Interpretations of Galileo's life and work are legion, but analyses of his *Letter to the Grand Duchess Christina* have largely revolved around two poles. One has been to understand its relation to the 1616 decree (historical analysis) and the other has concentrated on explaining how Galileo fended off theological criticism (literary analysis). Two of the most recent explanations are engagingly thorough. Finocchiaro's structural analysis sees no real internal tensions in Galileo's argumentation. In his view, Galileo was calling for the church to exercise a prudential patience which was best accomplished in an Augustinian mode: to recognize that the Scriptures were not intended to teach physical science, and to distinguish between those propositions which are demonstrable and those which are not. Galileo knew that he had no conclusive demonstration for the Copernican theory, but he also believed that the question of terrestrial motion was one capable of proof. The church would then be unwise in making judgments of its truthfulness until it had been conclusively demonstrated to be true or false. Finocchiaro's explanation has the advantage of seeing Galileo's hermeneutics as a seriously proffered method for resolving the perceived conflicts. This in turn allows us to see an inner coherence to Galileo's arguments, and also explains why Galileo probably did not think his method objectionable. To him, this hermeneutics was a conservative approach.[7]

McMullin's approach is more abstract but also more revealing of conceptual problems inherent in the *Letter*. His analysis emphasizes Galileo's indebtedness to Augustine by outlining five major principles, four of which he says Galileo culled from Augustine's *De Genesi ad litteram*.[8] McMullin also makes a plausible case that Galileo was no exegetical innovator in the *Letter to Christina* despite many scholars who still read Galileo's hermeneutics as historically novel.[9] Yet it would hardly have been to Galileo's advantage to offer a novel hermeneutics since he knew well the Tridentine requirement of patristic unanimity and the reluctance of many clergy to depart from well-attested precedents. Taking Galileo's language seriously demands that we see him intending to establish continuity with the patristic, especially the Augustinian, heritage.[10]

I argue that Galileo purposefully structured *three* approaches to the problem of reconciling Copernicanism and Scripture on Augustine's treatment of similar problems in *De Genesi ad Litteram* in order to establish his continuity with the patristic tradition. I also suggest that Galileo's reason for appealing to Augustine results in part from his view of the relativity of motion, an argument that reflects a closer continuity between his science and his interpretation of Scripture than has generally been recognized.[11]

The immediate context of Galileo's life that occasioned the *Letter to Christina* began in the year 1613 when he penned a letter to Benedetto Castelli regarding the issue of whether Copernicanism was compatible with Holy Scripture. This letter was later developed into a lengthier response and addressed to the Grand Duchess Christina, a letter written sometime early in 1615. Galileo became acutely aware of the theological dimensions of the Copernican debate through Castelli's letter of 14 December 1613 in which Castelli recounted to Galileo his extended conversation with the Medici family. Castelli reported that in response to the inquiry from the grand duchess concerning the compatibility of Copernicanism with Holy Scripture he "commenced to play the theologian with such assurance and dignity" that he won over the grand duke and his archduchess. Madame Christina "remained against me," says Castelli, "but from her manner I judged that she did this only to hear my replies."[12] This reluctance on the part of the grand duchess became the ostensible reason for Galileo's later letter when he had decided to enter the theological arena fully. Galileo's reply to

Castelli (dated 21 December 1613) was his first public expression of his views on Scripture. It is unlikely that he did any research in preparing this letter since it is dated only seven days after Castelli's. Yet in this reply Galileo says that Castelli's letter had given him "the occasion to go back to examine some general questions about the use of the Holy Scripture in disputes involving physical conclusions and some particular other ones about Joshua's passage."[13]

How much Galileo had seriously thought about scriptural interpretation prior to 1613 is not known. A comparison of his letters to Castelli with the more extensive one to Christina shows a greater development of thought and a more careful argumentation on behalf of his interpretation of the Bible. We know that Galileo was warned to stay clear of theological issues. Bellarmine had told Cesi that he considered Copernicus heretical. On 27 February 1615 Giovanni Ciampoli visited Cardinal Barberini, later Urban VIII, who told him "to caution Galileo to limit his arguments to mathematics and physics without getting into the theology of the matter."[14] Cesi had suggested that "the strategy should be that of preventing calumny against mathematicians, rather than supporting Copernicus as such lest more be lost than gained."[15]

Although Galileo was no theologian, his *Letter to Christina* is a masterpiece of theological reasoning. As we examine Galileo's obvious dependence on Augustine, only a close reading of both will show the depth of Galileo's reliance on the bishop of Hippo. As Augustine expounded the meaning of Genesis, he was compelled to deal with essentially the general problem facing Galileo, i.e. how scriptural interpretation relates to natural knowledge. Galileo's first method went beyond the strict issue of the theological acceptability of Copernicanism; he attempted to define the proper relation of science to theology in general. He contended that theology and the other sciences must be separate so that the integrity of each might be maintained.[16] Galileo refused to let biblical interpretation determine the content of the "lesser" sciences. His interpretation of the traditional phrase "theology as the queen of the sciences" was not one that made the other sciences inferior or that allowed exegesis to control science, but one in which theology and the other sciences complemented one another in displaying the glory of the Creator. Theology was queen because it "does deal with the loftiest divine contemplations and for this it does occupy the royal throne and command the highest authority."[17] His conception of theology grew out of his understanding of the purpose of the Scriptures, which was not to teach physical science but to show the way of salvation. Augustine too addressed this problem in the section on "the shape of the heavens" (*de figura coeli*) in book 2, chapter 9 of *De Genesi ad Litteram*. He answered the inquiry about what the Scriptures teach on the shape of the heavens by offering three different strategies. The first was to distinguish carefully questions of natural philosophy from questions of faith.[18] Augustine saw that the tendency to cite the Bible in answer to cosmological questions must be curbed in favor of recognizing the primary purpose of the Scriptures, namely, to instruct the faithful in matters of faith and morals. Galileo saw in Augustine's answer his first method for dealing with potential conflicts. In theology, Galileo could turn to no greater authority than Augustine, and he saw in his views the justification he needed to argue that the Copernican question was not a matter of faith.[19]

Galileo's second approach to handling conflicts between science and scriptural interpretation was to maintain that if one had demonstrative propositions in science, one could use these to reinterpret the Scriptures to accord with these demonstrated truths. Of one thing Galileo was certain: the Bible never contradicts physical conclusions that have necessary demonstrations.[20] According to this criterion, he must have in his possession a necessary demonstration of the Copernican theory, if he is to interpret the Bible correctly. During the years 1613 to 1616, Galileo repeatedly spoke of the necessary demonstrations and sense experience that proved the Copernican system beyond doubt.[21] His emphasis on necessary demonstrations is where scholarly difficulties begin, because it is clear that Galileo did not have in his possession the kinds of proof required by the common Aristotelian standards of his day. Furthermore, historians and philosophers are widely divided on what view of scientific knowledge Galileo actually held. Did he hold to the classic Aristotelian view of scientific knowledge consisting of demonstrative propositions, or did he adopt a more modern idea of the comparison of theories vis-à-vis evidence (relative superiority of theories)? Since Galileo never published a treatise on scientific knowledge per se, his views must be inferred from his work and his arguments for Copernicanism.

What no one doubts is that Galileo held to a realist view of mathematics as a vehicle for discovering natural truth. Galileo set for himself the task of searching out the "true constitution" (*vera constituzione*) of the universe. In his *Letters on Sunspots* (1613) Galileo openly expressed this view of science:

> These [hypotheses], however, are merely assumed by mathematical astronomers in order to facilitate their calculations. They are not retained by physical astronomers who, going beyond the demand that they somehow save the appearances, seek to investigate the true constitution of the universe—the most important and most admirable problem that there is. For such a constitution exists; it is unique, true, real and could not possibly be otherwise; and the greatness and nobility of this problem entitle it to be placed foremost among all questions capable of theoretical solution.[22]

Whether Galileo actually held an Aristotelian view of science, as his arguments in the *Letter to Christina* imply, or whether he adopted a more modest goal, his mathematical realism meant that Scripture might possibly come into conflict with the propositions of science. This possibility of conflict paralleled Augustine's situation and explains why Galileo modeled his second approach to reconciliation on Augustine's *De Genesi ad litteram* as well. In the next paragraph of Augustine's commentary, the ancient Father posed for himself the problem of reconciling the actual biblical language with the claim that the Scriptures were indifferent to the problem of the shape of the heavens that was implied in his first solution. The objection was that the Psalmist spoke of the heavens as spread out like a garment (Ps. 104:2). If that is true, then this statement contradicts those who say that the heavens are spherical. Augustine's response lays down the principle by which the interpreter of Scripture must hold to the authority of the biblical language if there is uncertainty about a physical question.[23] On the other hand, if there were indubitably demonstrated physical truths, then the interpreter was obliged to take these into account when explaining the biblical language. This second strategy seems completely unnecessary if Augustine believed in the separation of theology and science. Apparently, Galileo did not see the tension between these two approaches because his argument follows Augustine's exactly.[24] Yet, close attention to the language of both poses a more subtle question of how one knows whether the issue under debate—for Augustine the shape of the heavens, for Galileo the motion of the earth—is a matter of faith. Both Augustine and Galileo seemed to recognize that the strategy of separation cannot be used categorically to preclude any interpenetration between science and theology. This second strategy can be invoked when there is uncertainty over the status of a question.

Not content with two strategies for reconciliation, Galileo introduced a third which is even more surprising than the second. He seems to appeal to literal, physical interpretation. He argues that the sun rotates on its axis while being at the center of the universe. By its rotation the sun exerts a force on the other planets that causes their rotation. If the whole system of planets were to stop moving (including the sun), then the day on earth would be lengthened. Thus Joshua's words can be taken literally because the sun did indeed stop moving, which in turn caused the whole system to stop. In this way the biblical language on the Copernican theory is more literal than on the Ptolemaic.

This unexpected section of the *Letter to Christina* makes sense in the context of two important factors. First, Galileo's purpose in this third strategy was to cut through the argument that the Ptolemaic theory is closer to the teaching of Scripture than the Copernican theory is. There is nothing else in Galileo's work that would lead one to conclude that he seriously believed the Scriptures taught the Copernican theory any more than the Ptolemaic theory.[25] If Galileo believed that the Scriptures do not intend to give a theoretical account of the heavens, why does he attempt to reconcile the Copernican theory with the biblical words? It seems completely unnecessary if he believed in the separation of theology and science. It is precisely here that one is tempted to view the primary issue under debate as literal versus nonliteral interpretation since Galileo appears to have framed the debate in this fashion, but I will argue that this view of Galileo's third method is mistaken. It is more plausible to see Galileo as engaging in a form of hypothetical reasoning. Even if the interpreter insists on a literal reading of the Joshua story, such an interpretation still accords more closely with the Copernican system than with the Ptolemaic. In this way Galileo hoped to undercut the appeal of literal interpreters.

How and why did Galileo come up with this form of argumentation? The important factor is that Galileo is again depending on Augustine for this third strategy, as he has for the first two. In the third paragraph of the ninth chapter of *De Genesi ad litteram* (book 2), Augustine gives the same approach. He answers the objection of the imaginary interlocutor who argues that since the Scriptures speak of the heavens as a vault (Isa. 40:22), divine authority contradicts those who claim the heavens to be spherical. Augustine's answer is that even a literal interpretation does not contradict those who teach that the heavens are spherical if one recognizes the limited position of a terrestrial observer.[26]

I suggest that Galileo saw in Augustine's treatment a powerful argument for handling the problem. Like the great church Father, Galileo could argue that the Copernican theory did not contradict even a literal reading of the Bible. If his opponents argued that the church Fathers all agree on a literal interpretation of the Bible regarding the earth's motion (as Bellarmine and later Ingoli did), Galileo could respond in the fashion of Augustine that the Copernican theory is indeed consistent with a literal interpretation. Rather than being an unnecessary appendage, this portion of the *Letter to Christina* represents a type of argumentation that effectively supports Galileo's approach to the problem of biblical interpretation. Galileo's appeal is not so much to literal interpretation as it is to Augustine's authority.

In sum, Galileo's three methods are contradictory if each of them is taken to be seriously proffered hermeneutic approaches. However, if one sees the primary issue as the matter of historical continuity, then Galileo's adoption of Augustine's three methods of interpretation makes sense in the *Letter to Christina*. It becomes clear that Galileo has carefully constructed the three strategies to answer successively any possible objection he could encounter. His first method was the preferred one. He wished to separate strictly physical questions from theological ones, in which case the Bible would not be relevant to answering questions about the truth of the Copernican system. If, however, his opponents insisted that biblical statements are relevant in some way to our beliefs about the natural world and therefore cannot be strictly separated from physical science, Galileo introduced a second approach based on Augustine. This approach was to recognize the full authority of the Bible but also to invoke the principle of the agreement of (natural) philosophy and theology, an agreement based on the two books of nature and Scripture. Demonstrated truths in natural philosophy could be used in reinterpreting the Bible in a nonliteral fashion. Galileo's third method presupposes an interlocutor similar to Augustine's. If the interlocutor insisted that we have no reason to abandon the literal (physical) interpretation of the Bible for the point in question, both Augustine and Galileo answer that the proposed physical claim does not in fact contradict even a literal interpretation. In this way Galileo could believe there was no possible theological fault to be found in Copernicanism. This analysis of the *Letter to Christina* implies that the question of literal interpretation was only a consequence of a deeper issue, that of continuity with the patristic tradition. It is this continuity that Galileo wished to establish above all, and his three strategies, which appear contradictory on the surface, are reflections of a unity with Augustine. Galileo's method is not so much literal or nonliteral as it is historical.

Galileo's appeal to the patristic tradition was consistent with the principle enunciated by the Council of Trent. It was also consistent with Bellarmine's position expressed in his *Letter to Foscarini* in which the cardinal stated that one must proceed with the utmost caution in "giving to Scripture a sense contrary to the Holy Fathers and all the Latin and Greek commentators."[27] Galileo found that the church Fathers also noted accommodation in the biblical language. The Scriptures were addressed to the common man as he experienced the world. Thus Scripture speaks of God anthropomorphically as having eyes and ears. No one doubted that the Bible is accommodating itself to man's limited understanding when it uses such language. Based on the ancient hermeneutical tradition, Galileo argues that the Bible is also speaking of the heavens from an everyday vantage point. The sun appears to move; the earth does not. If God desired to teach the ancient Israelites of his miraculous intervention in nature, he would naturally describe the events of Joshua 10 as the sun stopping, since it appears to move in our everyday experience. In this way Galileo explained how the Bible can use language that differs from the way the heavens really are. The Bible is not wrong; it simply is not attempting to construct a theory of the heavens.

Recognizing Galileo's dependence on the patristic tradition does not explain, however, why he extracted from them an interpretative principle of accommodation. The belief in the necessity of dependence on the Fathers alone did not distinguish him from Bellarmine's position. There is an additional reason that does not apply to Bellarmine, namely, that the use of the accommodation principle has continuity with Galileo's principle of the relativity of motion. Galileo invoked no scientific principles in his *Letter to Christina* to argue for the mobility of the earth, but there is a striking similarity between his theological argument and his later scientific arguments at this point.

In the second day of the *Dialogue Concerning the Two Chief World Systems* (1632) Galileo attempts to answer objections against the diurnal motion of the earth based on the lack of observational evidence. Galileo's answer to the tower argument gave him an occasion to explain his principles of motion. The tower argument maintained that if the earth were moving, an object dropped from a tower would reach the ground not at the foot of the tower but considerably west of it. This clearly did not happen. Galileo's answer to this empirical objection invoked, among other things, the imperceptibility of shared motion, which assumed that the terrestrial observer moved with the earth.[28] Recognizing this implied that observations of motion (or lack of them) must always take into account the position of the observer.[29] This also implied that there was a position from which the earth's motion could be observed, a celestial position. The methodological principle of the relativity of the observer has its counterpart in the theological principle of accommodation. Galileo sought to explain the language of the Scriptures by invoking an interpretative principle that recognized the *relative position of the observer.* The lack of finding in the Scriptures references to a moving earth, Galileo maintained, was due to its divine Author (in a celestial position) adopting the observational position of the recipients. The function of observational relativity was also similar in both his science and his theology. In Galileo's science of motion, the recognition of the relative position of the observer was not, strictly speaking, a part of theory. It was a background assumption that was methodologically necessary. In theology, accommodation was not a doctrinal formulation; it functioned as a methodological assumption of the interpreter as he came to the text. Galileo no doubt felt justified in interpreting the language of the Bible as accommodating, since he already knew that the human observer is limited in his observation of the earth's motion. Just as no objection to terrestrial motion could be mounted if the imperceptibility of shared motion were taken into account, so no objection from Scripture could be adduced if accommodation were recognized.

My analysis suggests that Galileo did not attempt to prove that the Copernican theory was taught in the Bible; he only wanted to remove scriptural objections against it. Galileo believed that to attempt proofs of one physical system over another from Scripture would be as much a misuse of theology as of science. This sentiment is reflected in Galileo's note on his own copy of the *Dialogue:*

> Take note, theologians, that in your desire to make matters of faith out of propositions relating to the fixity of the sun and the earth you run the risk of eventually having to condemn as heretics those who would declare the earth to stand still and the sun to change position—eventually, I say, at such a time as it might be physically or logically proved that the earth moves and the sun stands still.[30]

Galileo saw more clearly than the theologians themselves how importing theology into physical disputes would wreak havoc on the church. The most important conclusion from my analysis is that Galileo's arguments were not motivated by a need to reinterpret the Bible in spite of the church's interpretation. Rather he understood his position to be historically what the Catholic church believed about how scientific questions should be settled, in that one of its greatest theologians had followed the same course. This conviction explains why Galileo could so boldly step into an arena that was not his, and why he may have seen himself as not entering into the theology of the Copernican question, a step he was sternly warned about. In his view, he was only clearing away faulty interpretations of the Scriptures. On many occasions Galileo professed belief in the unity of truth, and so it should not be surprising that his interpretation of the Scriptures and his science show a greater continuity with the past than has sometimes been recognized. More importantly, Galileo's diversity of exegetical approaches was motivated by a Catholic need, both his and his audience's, to demonstrate continuity with the patristic tradition, a strategy lacking in many of his Protestant counterparts.

POST-GALILEAN RESPONSES TO COPERNICANISM (FOSCARINI, INGOLI, CAMPANELLA)

Even before the decision of 5 March 1616 had been declared, Galileo's troubles provoked a number of strong responses, both positive and negative. Cardinal Robert Bellarmine received a letter from the Carmelite priest Paolo Foscarini that cast a long shadow over the issue of the earth's motion. Although Foscarini was a Copernican advocate, his underlying philosophy appears quite different from Galileo's. Galileo held that certainty of nature was possible through mathematical demonstrations whereas Foscarini sees uncertainty in every aspect of human reasoning. In his view, only divine revelation offers certainty, but God's revelation must be handled with the utmost care, lest it be distorted through well-intentioned but misguided applications.

Foscarini's *Letter* (dated 12 April 1615) contains a loose collection of theological and hermeneutical arguments, some of which do not always agree. He maintains that the purpose of Scripture was not philosophical because philosophy leaves the inquirer with controversy and uncertainty while the Scriptures were designed to foster the salvation of its readers by imparting certainty. For Foscarini, one cannot settle philosophical disputes by appeal to Scripture because of its higher purpose of lifting the human race above the vicissitudes of this world.[31] Foscarini maintained that "God has determined that only his holy faith is most certain, and everything else in this world is doubtful, uncertain, vacillating, ambiguous and two-sided."[32] It may seem contradictory to assert that the Scriptures are not concerned with natural philosophy and at the same time affirm that they are the supreme authority, but Foscarini reconciles these two assertions by making a clear distinction between matters that are *de fide* and those that are not. Natural philosophy is in the latter category; salvation, dogmas, and morals are in the former. This view follows closely the first strategy outlined by Augustine and adopted by Galileo. In the medieval theological heritage that Foscarini undoubtedly knew, such a distinction was commonplace. Where he

Epiftola

R. P. M.

PAULI ANTONII FOSCARINI
CARMELITANI,

Circa Pythagoricorum, & Copernici opinionem

DE MOBILITATE TERRÆ
ET STABILITATE SOLIS:

ET

DE NOVO SYSTEMATE SEU
CONSTITUTIONE MUNDI.

In qua SACRÆ SCRIPTURÆ *autoritates & Theologicæ Propofitiones, communiter adverfas hanc opinionem adductæ conciliantur.*

Ad Reverendiffimam P. M.

SEBASTIANUM FANTONUM,
Generalem Ordinis Carmelitani.

Ex Italicâ in Latinam Linguam perfpicue & fideliter nunc converfa.

Juxta editionem Neapoli typis excufam
Apud Lazarum Scorrigium Anno 1615.

Cum approbatione Theologorum.

Title page of Paolo Foscarini's *De Mobilitate Terrae* (1615). Courtesy of the Notre Dame Library.

differed from many medievals (e.g., Thomas Aquinas) was in his emphasis on the uncertainty of philosophical knowledge. Foscarini's skepticism may be explained by his formation in the Carmelite order. The reform of the order by Teresa of Avila and John of the Cross in the previous century stressed the uncertainty of all human knowledge and revived the ancient tradition of apophatic theology which acknowledged that even the highest expressions of human reason (theology) failed to capture the reality of God. On the other hand, Foscarini also asserts the second Augustinian strategy for reconciling Scripture with the results of science. This criterion applies when the results of investigation are true.[33] In the *Defense* of his *Letter* he says:

> From this it is clear that if the arguments of philosophy and mathematics have established a system of philosophy contrary to the Ptolemaic which has been accepted up to the present, we ought not to affirm emphatically that the sacred writings favor the Ptolemaic system or the Aristotelian opinion and thus create a crisis for the inviolable and most august sacred writings themselves. Rather we ought to interpret those writings in such a way as to make it clear to all that their truth is in no way contrary to the arguments and experiences of the human sciences (as Pererius says).[34]

As in Galileo's case, Foscarini gives every appearance of inconsistency unless one interprets his *Letter* as offering not so much a tightly knit defense of Copernicanism as a panoply of arguments for facilitating its acceptance. Perhaps Foscarini saw a distinction between absolute divine certainty and a limited human certainty that could be achieved in the realm of astronomy. In this case, it would be foolish and even harmful to use the Scriptures against what would be obvious to all. Foscarini no doubt hoped that his *Letter* would pave the way for the theological acceptability of heliocentrism, but Cardinal Bellarmine did not agree.

In many respects Bellarmine agreed more with Galileo's view on the matter of Scripture and demonstrations in science. In his reply to the Carmelite priest, the Cardinal argued that if the heliostatic theory had a "true demonstration," then "it would be necessary to proceed with great caution in explaining the passages of Scripture which seemed contrary, and we rather have to say that we did not understand them than to say that something was false which has been demonstrated."[35] Bellarmine seems to have left the door open for the acceptance of the Copernican system because he, Galileo, and Foscarini all inherited a belief in the unity of truth. Good science and good theology would be complementary, never contradictory. Cardinal Bellarmine, however, did not think such a demonstration possible. This was not only because he had not been shown one, as he explicitly says, but also because he thought Scripture would have been composed differently if a true demonstration of earth's motion were ever possible.[36] Bellarmine and Foscarini shared a skepticism about the role of mathematics in establishing certitude, while Galileo fully endorsed a mathematical realism. Bellarmine took astronomy (mathematics) to be at best a calculating device that would not give a true picture of the world although it might point to it obliquely.[37] For this reason, he stated that he always believed Copernicus to have spoken "suppositionally and not absolutely."[38] Natural philosophy and theology were another matter. In the cardinal's mind, they held a position of giving truth about the constitution of the world. For Foscarini, only theology and the revelation on which it was based was a certain avenue to truth.

Francesco Ingoli's *De Situ et quiete Terrae contra Copernici systema Disputatio* was published in 1618, three years after the proscription of the Copernican theory by the Congregation of the Index.[39] Ingoli explicitly and succinctly set forth three types of arguments against the Copernican theory. The order is significant. Since the questions under dispute were primarily astronomical, the mathematical problems of the Copernican theory had to be satisfied first, the major problem being that of parallax. The Copernican theory made the wrong predictions, according to Ingoli. Yet Galileo had taken the Copernican theory as a physical claim, and this suggested a second battery of physical tests that must be passed if this new theory was to be seriously entertained. Geokineticism, of course, was contrary to Aristotelian physical theory at the time. Finally, even if the requirements of astronomy and physics were met, the theological issues had still to be addressed to show that the theory was not contrary to good theology.

In a stair-step fashion, Ingoli argues that the Copernican theory does not satisfy the phenomena at all, since it fails in all three disciplines.

Ingoli seems to work on the assumption of the confluence of conclusions in all relevant disciplines and of the relevance of theology to all disciplines. Theologically, Ingoli argues that the *sensus litteralis* of the Bible ought to be maintained whenever possible; in his judgment that was true in this case as well. There was no reason to abandon the literal (physical) meaning of texts that teach the immobility of the earth, especially since convincing mathematical and physical arguments were not available. To bolster his interpretation, Ingoli appealed to the uniform interpretation of Joshua 10 among the Fathers:

> It is not sufficient to respond that Scripture is speaking according to our method of understanding. On the one hand, the rule to be used in interpreting Holy Scripture is to retain the literal sense when this is possible as is the case here. On the other hand, all the Fathers unanimously expound this place as meaning that the sun which moves stood still at Joshua's command. The Council of Trent (in its fourth session on the use of Scripture) abhors any interpretation which is contrary to the consensus of the Fathers. And though the Holy Council speaks on matters of morals and faith, yet it cannot be denied that an interpretation which is contrary to the consensus of the Fathers would displease those holy Fathers [of the Council].[40]

Ingoli also sees additional support for the immobility of the earth embedded within the tradition of the Church's liturgy. To some moderns this type of argumentation sounds strange until the principle of *lex orandi, lex credendi* (literally "the rule of prayer is the rule of faith") is recognized. Within the structure of Catholic theology, normative principles of faith may be derived from a deeply entrenched tradition of the Church's worship. This implies that if an article of faith can be found consistently in the history of the Church's prayers in the public liturgy, such an article may be considered *de fide*. This justificatory principle lies behind Ingoli's appeal to hymns and prayers of the *Divine Office*:

> The other argument is from the authority of the Church: for it sings in the vespers hymn of the third feria:

O great Creator of the globe
Who brought forth the world
by blowing movement of the water
You gave an immobile earth.[41]

Ingoli's appeal to the Liturgy of the Hours that was said by monks and priests shows how seriously he took this distinctively Catholic manner of ascertaining the theological acceptability of terrestrial motion. Taking Cardinal Bellarmine as his guide, Ingoli sees profound truths being taught in the nonbiblical sources of the Catholic faith. Ingoli did not see these extra-biblical sources as being contrary to the Bible, a charge often leveled against Catholics by Protestants, but as confirming the content of faith contained in both the Bible and oral tradition.

The thoroughness of Ingoli's theological treatment suggests that he may have thought of his work as an answer to the openly theological defense of Copernicanism by Foscarini. His language shows him well aware of the argument that the motion of the earth is not a matter of faith and morals. The quotation above ("though the Holy Council speaks on matters of morals and faith") is limiting, but it also indicates his commitment to settle the issue on the basis of exegesis and theology. He would probably not have done so had he not already been convinced of the astronomical and physical problems contained in the theory, but theology was clearly uppermost in his mind. For Ingoli, the decision of the Holy Office had been made. The only task for theologians was to find good arguments for the correctness of the decision, a task for which Ingoli called on all the available sources of Catholic doctrine. Ingoli's arguments failed, however, to address the problem underlying all the positions we have examined so far, that of deciding what to extract from the church Fathers. He gives no reason for rejecting the language of accommodation in favor of a literal reading, nor any authoritative basis for knowing when the *sensus litteralis* should be retained. He did not extract his assertions from the text of the Council of Trent itself because the Council did not directly address that issue. His document shows clearly, however, that the issue of continuity with the patristic tradition was always an assumption of the discussions in Catholic Europe.

In 1622 there appeared in Frankfurt another defense of the Copernican position by a Dominican friar, Thomas Campanella, who had already encountered various difficulties with authorities in Rome.[42] Although it is impossible to know when Campanella wrote his *Apologia pro Galileo*, Femiano and Blackwell have argued strongly for a date prior to the condemnation of Copernicanism (5 March 1616). Campanella's own indications are that this document was requested by Cardinal Caetani as a theological report on the question of heliocentrism.[43] Like other defenses of the new astronomy, the *Apologia pro Galileo* assumed a distinction between a *de fide* doctrinal issue and a philosophical theory. Campanella maintained that such a distinction is necessary to avoid the scandal of the church being proved wrong in the future, should Galileo prove the Copernican theory right.

Campanella included much stronger claims for Copernicanism than Galileo. Campanella saw a closer fit between Scripture and the Copernican theory than with any previous natural philosophy because Galileo's work revealed a new system of the world that would be the foundation of a total philosophy. One such argument concerned the material of the heavens. Some of his objectors argued that belief in the corruptibility of the heavens contradicted Scripture because God was perfect and heaven was the dwelling place of God. Campanella vigorously argues that the corruptibility of the heavens is not contrary to Scripture or the Catholic Faith; he takes the discoveries of recent decades as sufficient proof of the material (and therefore corruptible) substance: "sunspots and new stars in the starry heavens, and comets above the moon clearly show that the stars are other world systems."[44]

For Campanella the varied and tortuous interpretations of the Fathers regarding the firmament had been disproved by Galileo's discoveries. Those interpretations were based on various ancient Greek philosophers (e.g., Empedocles, Plato, Aristotle) and attempted to reconcile the text of Genesis 1 with the philosopher's views of the nature of the heavens. Galileo's work allows the Scriptures to be interpreted properly and shown to be true:

> Galileo has shown that there are mountains on the moon, and Scripture agrees with him because Genesis 49 [49:25] and Deuteronomy 33 [33:13–15] speak of fruits and mountains and hills in the heavenly bodies.

Therefore, Sacred Scripture, taken in its literal sense in all its passages, agrees only with Empedocles, and it agrees with the others only in a mystical sense or by doing it violence [*violenter*]. But Empedocles was a Pythagorean, as also is Galileo. So Galileo ought to be praised. After so many centuries he vindicates [*vindicat*] the Scriptures from ridicule and distortion by observable evidence [*per sensatas experientias*]. And he has shown that the wise men of this world are fools, and Sacred Scripture should not yield to [*obsequens*] them, as has happened up to now, but they ought to yield to Sacred Scripture. And this does not degrade our planet. For humans will be raised up with Christ, their leader, above the stars and above all the heavens. So from this it is clear that we are better off than those wise men.[45]

For Campanella the Copernican theory becomes not only a possible, or even a certain, theory distinct from Scripture; it becomes the best tool for reinforcing the authority of Scripture interpreted in a physical manner. Yet Campanella's enthusiasm for Galileo did not rest on the Copernican theory alone but on the entire panoply of discoveries made in recent decades. He viewed Copernicanism not only as an issue of terrestrial motion but as a harbinger of new philosophy that would confirm the truth of the Scriptures. Campanella's willingness to invoke the literal physical sense of biblical words shows that even zealous Copernicans could interpret the Scriptures in a very physical manner. His approach to the theology of Copernicanism can be explained best by his background in the Dominican order. With a thorough exposure to Thomas Aquinas's theology, Campanella no doubt considered himself following in the Angelic Doctor's footsteps. Aquinas attempted to synthesize the best of Aristotelian natural philosophy with the content of the Catholic faith. Since many of the specific aspects of natural philosophy had changed since Aquinas's day, Campanella thought that incorporating those changes into a unified philosophy would do for the present what Thomas's work had done for the thirteenth century. The Copernicans were by no means united in their approach to the problem of Scripture and natural truths. Foscarini, Galileo, and Campanella, all ardent Copernicans, exhibit slightly different approaches to the problem of reconciliation.

COSMOLOGY AND INTERPRETATION IN
CATHOLIC EUROPE

Earlier I suggested that the theological context of Catholic Europe shaped the parameters of discussion of Copernicanism in the seventeenth century. The individual discussants treated above were clearly working within a framework set by the Council of Trent, but many of the dictates promulgated by the council were deeply rooted in earlier Christian thought. What the council did not address directly, of course, were issues in natural philosophy, and it was precisely there that serious differences arose as to how inquiry into nature related to the theological positions the council espoused. What no one doubted in these disputes was the doctrine of the unity of truth. When Galileo wrote to Castelli that it was impossible for two truths to conflict, he was expressing nothing less than a deeply embedded tradition that reached back to the earliest conflicts of the church with heresy. This belief was no untested assumption, for it came with a vengeance in the thirteenth century with Siger of Brabant and the Averroists. As best can be determined, these scholastic philosophers maintained that certain truths derived from Aristotle could be demonstrated as true yet they were also in conflict with the Christian faith. While Siger added that these truths must yield to the truths of the faith, his method raised problems for the notion inherited from Augustine that God, the source of all particular truths, would not allow contradictions. Thomas Aquinas addressed this doctrine of double truth and reaffirmed the ancient Augustinian doctrine that the unity of the human intellect demanded noncontradictory relations among particular truths. By the Council of Trent in the sixteenth century, there was no dispute about the necessity of reconciling apparently conflicting truths.

Just as deeply ensconced in the Christian tradition was the separation of disciplines that is observable in Galileo, Foscarini, and Campanella, as well as in Ingoli. We already saw in the Protestant writers this recognition for keeping different types of information and truth claims separate. Each discipline, whether astronomy, physics, or theology, must appeal to its own standards of adjudication for resolution of conflicting claims. This relative separation of sources of truth lies behind the warnings to Galileo by Bellarmine and others to stay out of the theological arena. And there seems little doubt that the proscription of Copernicanism would not have taken place had Galileo remained silent about the theological dimensions of terrestrial motion. The well-meaning but ultimately harmful contributions of Foscarini and Campanella could only heighten the perception that a moving earth was harmful to Christian truth and the Catholic faithful.

Looming large in the Catholic world was the issue of patristic unanimity. We observed among Protestant disputes a decreasing emphasis on continuity with the church Fathers as the divide between Rome and the Reformation deepened. Even at the outset of the Reformation (e.g., Luther, Melanchthon, Calvin) a selective use of the patristic witness can be observed. The Catholic Copernicans had a deeper obligation to query the Fathers on the matter of the earth's motion, and their answer required a greater subtlety of interpretation than was necessary among Protestants. A comparison of Galileo, Foscarini, and Campanella with Ingoli in their respective treatments of the Fathers signals how differently both sides of the Copernican debate read the silence of the Fathers. Most Copernicans drew on the Augustinian tradition and argued for a separation between scientific and theological questions. The differences between them and their opponents do not lie in different views of authority, since the Copernicans also acknowledged the inviolable truth of Scripture and the place of the church as the supreme interpreter of Scripture. Rather, the Copernicans argued that heliocentrism was not a *de fide* matter, and that Scripture and tradition cannot be expected to address or settle the issue. The silence of the Fathers was taken as evidence of this proposed separation.

Campanella attempted to show that the Fathers of the church had not in fact delivered authoritatively or uniformly on this matter. Perhaps because Foscarini's *Letter* had failed to address the question of patristic precedent, Campanella culled quotations from the Fathers and the medieval scholastics, preeminently Thomas Aquinas. One of the greatest of the Greek Fathers, John Chrysostom, confessed to not knowing whether the earth moves. He was supported by a whole host of other patristic witnesses. Campanella therefore concluded that the authorities in Rome could only condemn Galileo because they did not understand the Fathers.[46] The non-Copernican Ingoli searched for the consensus of the Fathers on any given issue and, whenever he could discover it, he took their consensus as an indication that the issue in question was a matter of faith and morals. His appeal to the unanimous consent of the Fathers pointed to the lack of reference to a moving earth. The silence of the majority of the Fathers on this issue was taken as nonapproval. Ingoli's method moved from historical precedent to the present question in dispute. For him, the motion of the earth, while not prima facie a theological question, became one by virtue of the Tridentine criterion.[47] The Fathers may not have delivered extensively on the motion of the earth but that only showed in Ingoli's judgment that they did not or would not approve the notion of a moving earth. This kind of extensive investigation into the Fathers would not have been as necessary or sufficient in a Protestant context, especially in the Reformed Netherlands where the principle of *Sola Scriptura* implied the nonbinding character of ecclesiastical synods.

The controversy between Catholic Copernicans and non-Copernicans hinged as much on different interpretations of the astronomy as it did on hermeneutical approaches. Like many in Europe who held the time-honored view of astronomy as saving the appearances, Cardinal Bellarmine advised Foscarini not to take astronomical theories as a window on truth but as an instrumental claim. As an instrument, no theory in any science could be contrary to Scripture. As many historians have noted, the revolutionary character of Copernicanism was not only a reversal of the place of sun and earth but entailed the realist claim made by Kepler, Galileo, and others that mathematical formulas were direct reflections of the true constitution of the universe. When the Copernicans argued for heliocentrism in a realist fashion, the Catholic opponents invoked the literal (physical) meaning of Scripture to argue against the theological acceptability of a moving earth. Yet even an appeal to a physical meaning of the biblical words did not settle the issue because Galileo's and Foscarini's methods of argumentation show that they too used literal interpretation. They used the *sensus litteralis* in the historical sense of the biblical text teaching what the authors and recipients of ancient times understood.

An emphasis on the *sensus litteralis* in the Catholic debates was not surprising for two reasons: one historical, the other contemporary. Whenever hermeneutical controversy arose in the history of the church, a renewed stress fell on the literal sense because there was a need to gain control over varying interpretations. The second emphasis emerged from the controversies of the Reformation itself. The Reformed interpretation of the Eucharist took Christ's words of consecration (*hoc est corpus meum*) in a less than physical manner. From the vantage point of Catholic dogma, this looser reading minimized the authority of the Bible. Although the Protestants had emphasized the centrality of the Bible, their interpretations denied its authority. Yet the same forces that drove Catholic theologians back to the *sensus litteralis* also impelled Protestants. The Protestants were just as sure that the Catholics had missed the plain meaning of Scripture in many of their impious doctrines. The Protestants had many reasons to emphasize the literal meaning of the text because, in their view, this alone would lead to a proper understanding of Scripture. In fact, the *sensus litteralis* was sometimes the only common ground that disputants had. When the subject became the theological implications of natural philosophy, it was quite natural for both sides to appeal to the *sensus litteralis* to settle the dispute. Both Catholic and Protestant disputants recognized accommodated language in the Bible, but a predisposition to emphasize the *sensus litteralis* predominated because of other theological issues that had demanded such an emphasis.

Interpreting the History of Early Modern Cosmology and the Bible

*T*he early moderns used a curious mixture of common ideas in different ways when they sought to read the heavens above and the Book below. What they all had in common was as important as what divided them, but those commonalities are more difficult to see in the midst of controversy. Catholics and Protestants, Lutheran and Reformed alike, shared the tradition of the two books. The obligation to investigate and understand the Book of Nature and the Book of Scripture was shared by all because it was so deeply embedded in their historical consciousness. All the actors in this drama would have agreed with Hugh of St. Victor's statement of this tradition in the twelfth century, even if they had never read it directly:

> For the whole sensible world is like a book, as it were, written by the hand of God, that is to say, created by divine power, and each of its creatures are like forms, devised not by human effort, but rather established by the divine will in order to make manifest the wisdom of the invisible things of God.[1]

A belief that the Book of Nature revealed the Creator reached of course back to St. Paul's words in the Letter to the Romans (1:20), "the invisible things of God are known through the things that are made." This statement in turn depended on those texts such as Psalm 19:1 that were quoted repeatedly by writers we have investigated. That they should have spoken of nature as a book is significant. Their Christian heritage had imbued them with a sense of the authority of the Bible as revelation, and the two-books tradition had allowed them to theorize about nature because it also was an instrument of divine revelation. That tradition, however, gave little guidance as to how the two books should relate to one another or how conflicts between readings should be resolved. That open-endedness allowed them to debate new cosmologies and natural philosophies, but it also required them to mesh new views of nature with adequate interpretations of the Bible. The more recent historiography of the scientific revolution has unveiled a world of greater diversity in natural pursuits, and my research has shown a much wider diversity of hermeneutical strategies in this process of reconciliation.

The hundred or so years between the appearance of Copernicus's masterpiece, *De Revolutionibus*, and the debates among the Dutch saw many different approaches to the problem of reconciliation, but never one with a rational Copernican science on the one side and benighted medieval biblicism on the other. The Bible and science were employed and embraced on both sides of the Copernican question, and many similarities of epistemology and hermeneutics cross over that divide. Indeed, the complexities of this history suggest that the best way to capture the divisions may not be Copernican/anti-Copernican at all. A far more profound division was the realism/antirealism arguments, but even here one finds gradations rather than rigidities. Biblical interpretation did not become relevant to heliocentrism without a prior commitment to realism in astronomy and physics. Copernican advocates (e.g., Kepler, Lansbergen) and opponents (e.g., Froimond, Ingoli) shared this common commitment to realism, either of a purely mathematical type or a convergent realism that incorporated physical principles that went beyond mathematics. Most realists shared a belief in biblical relevance at least at some level. The question was not so much *whether* the Bible was relevant but *how* its texts and/or dogmas played into the true view of the universe that these astronomers were searching for.

If it appears that theology became more limited in the early modern period by having to stand alongside other disciplines in interpreting nature, this was only because any discipline has to make room for new-comers when those newer ones make bold and well-warranted claims. During this period the mechanical and chemical philosophies con-tended against one another as alternative philosophies of nature that would open up the secrets of the cosmos. Eventually, however, the pri-mary disciplines of each, physics and chemistry respectively, had to complement one another as different levels of description. Neither one proved to be *the* key to the whole of nature. Similarly, theology came to be only one among the many disciplines that would contribute to the understanding of nature. This was not a process of eclipsing theology by other superior disciplines, à la Draper and White, but an inevitable consequence of two beliefs shared by virtually all early moderns. They believed that truth could be discovered through rational and empirical investigation as well as through divine revelation, and they also believed that all truths of nature, no matter their source, must be reconcilable with one another. We have seen that they adopted an amazing variety of strategies of reconciliation, but they all were founded on the Augus-tinian dictum of the unity of truth. In this respect, Augustine turns out to be not only the most influential theologian of the West in strictly theological issues but also the greatest Christian Father to have shaped discussions of natural philosophy and theology.

Mathematical Realism and Disciplinary Boundaries

The diversity described in earlier chapters should not blind us to the deeper commonalities held by the historical participants we have treated in this story. Although it is precarious to generalize about all Copernicans, for example, we shall miss something crucial to their work and opinions if we do not recognize the underlying assumption of *mathematical realism in astronomy*. The transformation that oc-curred in the goals of astronomy from saving the phenomena to ac-counting for the real cosmic system was so profound that it eclipses the issue of the moving earth in importance. Without the adoption of a re-alistic goal for astronomy, the question of a moving earth would have never created the furor it did. Realism of course existed in medieval natural philosophy, but it was the transference of this aim of finding natural truth to astronomy that was *the* fundamental revolution in the sixteenth century. That transference was grounded of course in a confidence in the language of mathematics as *the* proper vehicle for truth-telling about the heavens. Historians have sought to ground this mathematical confidence in earlier philosophical traditions, preeminently in the revival of Neo-Pythagoreanism, but what is more interesting than its pedigree is its purpose. It is not accidental that Kepler and Galileo invoked the metaphor of "language" to describe mathematics as the proper tool for understanding the Book of Nature. The one who properly interpreted the Book of Scripture must also understand its language, a task that went beyond grammatical knowledge of ancient tongues. The linguistically competent exegete of Scripture had to pierce beneath the phenomena of the Bible to elucidate the divine Author's intention. In a similar fashion, Kepler, Galileo and others who adopted this viewpoint were appealing to a knowledge of mathematics not only to describe the phenomena of nature but to pierce to its very core and to see the divine Author's intention in so constructing the universe. Perhaps more than any other, Kepler believed that he had unveiled the mind of God.

The issue of realism yields an explanation for the relevance of theology to terrestrial motion, celestial matter, and the manifestation of the divine plan in the physical universe. Few astronomers dared to assert openly in the 1540s what Rothmann and Galileo would later argue, namely, that mathematics was the key to God's design of the universe and that therefore mathematical astronomy was sufficient to capture the true world system. The instrumentalist interpretation pursued by Erasmus Reinhold and Gemma Frisius, and surviving much later in Robert Bellarmine, was not so much a conscious decision to limit astronomy as it was an assumed premiss of their work, a presupposition that had the felicitous endorsement of historical precedent. However, by the 1580s the questions among astronomers and natural philosophers had changed into a search for a world system that met the criteria of mathematical prediction, physical theory, and cosmological harmony, a shift in which hermeneutical and theological questions were also considered relevant. This growing sense of realism required that prediction and mathematical models be increasingly subordinated to the larger goal of mapping the entire cosmos, a goal shared by such diverse figures as Rothmann in Germany, Brahe in Denmark, and Clavius in Rome, though their commitments to different systems deeply divided them.

The medieval division of labor, in which technical astronomy was distinct from cosmological inquiry, was rooted in Aristotle's works (e.g., *De Caelo*) and was explicitly stated by Simplicius in his commentary on the *Physics*. Although this work was not known in the Middle Ages, the substance of that division of labor was embraced and practiced.[2] By the fifteenth century "the *Theorica* compromise" emerged that came to expression in Peurbach's *Theoricae Novae Planetarum*, an attempt to unite Ptolemaic hypotheses into a physical system with Aristotelian concentric spheres. Methodologically, it represented an effort to overcome a rigid division between astronomy and cosmology. These attempts had very little effect on the educational texts of the sixteenth century, where the separation of these disciplines was paramount.[3] In ongoing research, however, the lines between technical mathematical astronomy and natural philosophy began to blur, and a forthright mathematical realism emerged in figures like Rothmann, Kepler, and Galileo. Kepler was the first to proffer an explicitly physical version of heliocentrism, but he was not the pioneer of realism in the early modern era; Tycho himself expressed a similar goal many times before Kepler's *Mysterium Cosmographicum* appeared (1596). It was this physical realism that accounts for the rejection of heliocentrism in the latter half of the sixteenth century and early decades of the seventeenth. Yet, as increasingly better physical versions of a sun-centered universe were presented, the plausibility of the Copernican system grew dramatically. The plausibility was enhanced by the perceived coherence of Copernicus's system, a point emphasized by Kepler and Lansbergen.

Westman's analysis of the disciplinary status of astronomy and Jardine's analysis of the issue of realism have resulted in a much finer brush-stroke being applied to early modern thought. Jardine painted the positions on realism as a three-way distinction between realists (Clavius, Ptolemy, Copernicus, Kepler), skeptics (Pontano, Ramus, Frischlin, Baer) and skeptical realists (Fracostoro, Amico), although there are many gradations and nuances in these figures.[4] Nicholas

Reimars Baer, for example, seems to have denied any possibility of knowing the true system but Peter Ramus fully endorsed the pursuit of the true system; he only denied that astronomy with its hypotheses could reveal that system. Underlying Jardine's typology were several questions, the most prominent of which was, "Did the historical figure believe in the possibility of knowing the true system and/or causes of planetary movement?" His analysis has the considerable advantage of allowing for an intermediate answer, namely, that among the many hypotheses invoked by the astronomer, one set corresponds to the true world system. I argue that Tycho falls into this intermediate category of the skeptical realist. However, Jardine's classification does not answer another important issue that the early moderns faced—the question of whether one discipline (e.g., astronomy) is the key to the true world system or whether all disciplines must be taken into account. The problem facing them was not only the status of hypotheses and the possibility of causal knowledge but the relevance of types of information.

We have observed that theology was often thought relevant in the early modern era, not because it was imposed on astronomers by theologians from the outside, but because the astronomers and natural philosophers themselves believed that the Bible and theology had much to say about the heavens. This was often as true for Copernicans as it was for non-Copernicans. The Bible became relevant for Kepler, Lansbergen, and other Copernicans because they were anxious to invoke theological concepts in their meditations on the heavens. Previous scholarly discussions that examined only the Copernicans' arguments about the motion of the earth have failed to account for a fuller range of citations and quotations that not only show they considered the Bible relevant but also that they held to a kind of theological dimension to their science. Kepler, Lansbergen, and Galileo all spoke of portraying the glory of God in their work by demonstrating the precision and grandeur of the heavens. In their view, some of the truths in the sacred page are also to be found in the celestial pages if only one learns the language in which those truths are written. The dynamics of Protestant-Catholic polemics informs some of these arguments. Galileo referred to the heretics' embracing of Copernicanism, wanting to make sure that the Catholic world did not lag behind in its pursuit of the new science. Kepler's and Lansbergen's use of the Trinity sphere-icon did not correlate, of course, with any specific Protestant doctrine, but it did serve to buttress their claim that they were as orthodox in their science as they were in their theology. Perhaps both Catholics and Protestants hoped that showing theological truths through their science would reinforce the truth of their theologies.

When the issues of terrestrial motion and cosmological reality were forthrightly faced, the problem of *disciplinary hierarchy* arose. The question of which discipline stood higher in the hierarchy required a prior answer as to what kind of information a specific discipline provided. The traditional hierarchy saw astronomy as providing the mathematical models and physics as selecting among those models the one or a few that met the demands of Aristotelian physical principles. Theology added the final component by acting as a guard against physical models that violated theological principles. While theology served to limit objectionable models, it was rarely, if ever, seen as the primary discipline in cosmology. This type of hierarchy required that the higher discipline must explain the principles of the lower, but the lower had no possibility of explaining the higher. Thus, any implied motion in astronomy must be derivable from physical principles. This conception of disciplinary hierarchy also involved a kind of downward negation. The higher disciplines had the possibility of negating claims from the lower discipline but not vice versa, a situation reflected in the criticism advanced against Tycho's system that physical principles did not allow for more than one center of motion. Theology still stood, however, in a somewhat distant position from physics and astronomy. The planetary motions of astronomy must be derived from or at least be closely related to the principles of motion in physics, but the power of theology to negate physics was more limited. Even such ardent Ptolemaic defenders as Froimond and Ingoli acknowledged that theology had no principles of motion per se.

The disputants in the Copernican debates were careful not to confuse the sources of truths to be derived. They all held some notion of separate disciplines, but they ordered them into different hierarchies. One important difference was *unidisciplinary vs. multidisciplinary approaches*. Christoph Rothmann argued strongly that astronomy alone could yield the true system, but this was in fact a minority position at the time. His high confidence in mathematics was shared by most, especially Kepler and Galileo, but he denied the relevance of physics and theology. This exclusivist position was not endorsed by Kepler. Kepler's view required the elaboration of physical principles to account for the mathematical precision and elegance that God placed in the universe. He sided with Tycho's view that all three disciplines had positive roles to play in the final world system. Rothmann's was a pure *mathematical realism.* Kepler's and Brahe's was a *convergent realism* that required all three disciplines to agree in their conclusions. This distinction refines Jardine's analysis and implies the need for cross-categorization. Compared to Kepler's classification there can be no doubt that Brahe's is more difficult. Tycho and Kepler had very different views of how the true system would be found, of how disciplinary convergence came about. Both astronomers believed that the universe embodies divine presence and meaning, but they saw that embodiment in different ways that affected their views of disciplinary interaction.

Tycho often expressed confidence in finding the true system, but he also tended to view hypotheses as convenient fictions. I suggest that Tycho was in fact a skeptical realist, believing that knowledge of the true system was possible but not in the a priori manner of Kepler. Tycho held that the principles in each discipline were distinct. Ancillary disciplines might throw light on the question posed, but each discipline had to bear its own weight in addressing questions proper to its own sphere. The motion of the earth could only be properly answered by physics. The Bible held a special status simply because it was divine revelation. Consequently, although it is not a book of physics, it may contain physical truths in passing. But the Bible did indicate that features in the universe have a meaning beyond their mathematical and physical parameters, a belief that lay at the heart of Tycho's astrology. Mathematical astronomy and physics would never yield astrological meanings just as certainly as the Bible would never yield a physical philosophy. Truths in one discipline would not demand a corresponding principle in another, although all disciplines had to agree in their conclusions on a particular issue. Tycho held to distinct lines of differentiation.

Kepler's very different view of disciplinary interaction grew out of his apriorism, in which the theological content of the universe is ipso facto in the creation. Underlying truths in all disciplines are the same, but there are mathematical, physical, and theological expressions of those truths. This belief allowed Kepler to embrace a methodology that Tycho would never have endorsed. For Kepler, if a statement or implication were true in one discipline, one could know a priori that it must be true in all others. The theological truths expressed in the Bible had mathematical and physical reflections so that heliocentrism was not only mathematically but theologically necessary. Kepler's treatment of the Bible may be misleading in this regard because he appears to say on the surface that the Bible does not contain astronomy. I argue that Kepler's view of the Bible and scientific disciplines must be understood on two distinct levels, one exegetical and the other theological. He insisted that the Bible is teaching not scientific principles but redemptive truths. Proper interpretation does not seek physical truths in Scripture. But on a deeper level the theological truths expressed in the Bible (Trinity, harmony in the church) are also expressed in nature. On one level, Kepler has a methodological separation but on another a theological unity. These differences between Tycho and Kepler show why it is perilous to argue for a Copernican view of disciplines on the one side and a traditional (non-Copernican) view on the other. Both held to an essential agreement of disciplinary conclusions, but they had different approaches as to how those conclusions would be reached.

Both Tycho's and Kepler's views of disciplines bear the earmarks of Lutheranism because both display a certain *centripetal* force characteristic of the German Reformation. Both astronomers were nurtured in a religious and social environment in which church and secular leaders attempted to overcome existing political divisions and theological acrimony. The initial separation from Rome resulted in the splintering of ecclesiastical regions of Germany by the 1560s, and this phenomenon demanded an ardent search for unity among Luther's followers. The theological animosities sketched in chapter 2 eventuated in the Formula of Concord (1580), a decisive turn toward Lutheran unity. This drive toward unity after the initial break with Rome, however, was embedded in the theology of Lutheranism itself. The Lutherans viewed themselves not as an alternative church to Rome but as a reform movement within the catholic faith. Their doctrine of the Eucharist, too, reflected the call to unity since it required belief in the real presence as a precondition for reception, a practice that excluded the Calvinists. The Lutherans had an antagonism to papal power, to be sure, but their hottest wrath was saved for the Calvinists, whose views of the Eucharist and the church signaled a much more serious departure from the ancient catholic faith in Lutheran eyes. I suggest that although Tycho and Kepler both expressed Calvinist leanings at times, their belief in the necessity of converging truth was fostered by the Lutheran context in which they lived. This was not the case in the Reformed Netherlands. The strongest separation arguments, we have seen, were to be found among the Dutch Calvinists. Their theology, church structures, and social realities had a *centrifugal* force, a decentralizing tendency that emerged from the struggle to free themselves from Spanish rule and Roman ecclesiastical domination. From the very beginning, Reformed theologians like Calvin argued against linking the biblical faith too closely with a specific natural philosophy. Lansbergen and Wittich, both professional theologians, invoked biblical and theological reasons for not expecting a cosmological contribution from exegesis. The Reformed view of the church differed from the Lutheran, and the break with Rome was more severe, entailing a choice between allegiance to "Popish" religion or to the true Reformed faith. The church did not need to be reformed from within but rediscovered in Geneva and its offspring. These tendencies are reflected in the Dutch separation of Cartesian philosophy from Calvinist theology, but it was not Descartes' philosophy that created such openness. Rather, it was the separatist tendencies of Reformed thought that paved the way for the reception of Cartesianism, not universally or inevitably, but extensively and pervasively.

An interesting transformation took place between Melanchthon's separation-of-disciplines argument in the 1540s and that of the Dutch Cartesians in the 1640s. The Lutheran humanist wanted to keep astronomy, physics, and theology distinct because mathematical astronomy had no contribution to make to cosmology other than giving hypotheses. He might have endorsed the *Theorica* compromise but he certainly would never have embraced Rothmann's *sola mathematica* nor Kepler's Neoplatonic unification. Physics and theology would unite their contributions to lead to cosmological truth. Kepler's transformation was to unite all three disciplines, expecting all to reflect the same modified Copernican system. Others such as Tycho and Clavius wanted the same union of disciplines but with very different conclusions. By the fourth and fifth decades of the seventeenth century, the separationist arguments were beginning to win the day, a development that probably accounts for the demise of Kepler's cosmology while his astronomy continued to exercise considerable influence. The Cartesian Copernicans in the Netherlands shared Kepler's confidence in mathematics but argued for disciplinary separation because they feared a loss of freedom for philosophizing and a violation of exegetical integrity if the Bible were pressed into cosmological service. From Melanchthon to the Cartesians, the separation among disciplines was transformed from a hierarchical and convergent concept into a true and lasting division, a development that was fostered by Reformed theology.

EXEGETICAL PRINCIPLES IN COSMOLOGY

Disciplinary boundaries would have been a problem even if the early moderns had never considered theology, but in fact their views of how the disciplines should interact were related to their conceptions of hermeneutics. Those who viewed the Bible as teaching scientific truths thought that its words and phrases had to be incorporated into a total cosmology. Those who thought of the Bible as a book of faith without scientific content (e.g., Rothmann) did not feel compelled to integrate biblical language into their natural-philosophical theories. Scholarly discussions of biblical exegesis have distinguished between literal and nonliteral approaches, the former being urged by opponents of terrestrial mobility and the latter by Copernicans. Some scholars refined this distinction with the recognition that many Copernicans appealed to accommodation in the Bible, a move that did not so much deny literal interpretation as it denied the relevance of biblical language to natural inquiry. The discussion above made it clear that questions of relevance were based on prior judgments of the subject matter of each discipline. Now we must focus on how exegetical judgments affected views of the relevance of theology.

The unmistakable diversity of exegetical approaches discovered in earlier chapters undermines the simplistic dichotomy between literal vs. figurative hermeneutics, but this diversity existed prior to the advent of the Copernican debates. In fact, all the exegetical judgments made by these figures had precedents in the earlier history of Christian theology. There were not many manuals of interpretation written in the ancient Christian church, but the most influential, Augustine's *De Doctrina Christiana*, shaped hermeneutical theory and practice for centuries.[5] Here and elsewhere Augustine displayed greater flexibility in what was included in the *sensus litteralis* than historians of science have recognized. He often glossed the *sensus litteralis* as a *sensus historicus*, the meaning intended by the biblical author that should be interpreted according to the rules of grammar and the facts of history of the author's time. The notion of the author's intention became central to subsequent hermeneutical theory, but it was complicated in the case of the Bible because there was also a divine author behind the human words. Was the divine author's intention identical with that of the human author? If not, how would one discern God's intention amid the human author's words? As we saw in chapter 1, the answers to these questions yielded several approaches in the hexaemeral literature of late Christian antiquity that bore on natural philosophy. Ambrose of Milan, Augustine's mentor, read the text of Genesis as a straightforward physical description of nature without much use of metaphor or accommodation. The six days of creation were taken as natural days.[6] Augustine, on the other hand, had a much more subtle view of the hexaemeral language. He maintained that the days of Genesis were adjusted to our reckoning of time because the totality of the created order was created simultaneously at the beginning (Gen. 1:1). Further, Augustine warned against invoking Scripture too quickly to refute positions in natural philosophy lest the ignorance of Christians put off those who have solid knowledge.[7] These two approaches formed the parameters of interpretation well into the early modern period, the one treating literal interpretation as physical descriptions comparable to those in the special sciences, the other taking biblical language as culturally determined and as having fewer connections to natural philosophy.

Both Ambrose's and Augustine's approaches to biblical language were subsumed under the *sensus litteralis*, a concept that was not a rigid category, either in theory or application. Augustine sometimes took the Bible as giving a straightforward physical account, and Ambrose also recognized figurative and metaphorical uses of language. The *sensus litteralis* did not allow two different approaches to interpretation, but it did contain within itself two shades of meaning, one of which is most commonly adopted by historians in describing the defense or attack of terrestrial motion. The first is the physical meaning of the term in which the text is taken to describe a physical, cosmological reality in a rather straightforward, almost iconic manner. The second commonly understood sense allows for figurative meanings and accommodated speech. In the traditional fourfold sense of the Bible (literal, allegorical, anagogical, tropological), *sensus litteralis* is that which is conveyed by the historical-cultural context in which the words were originally written. The correspondence between the words in the text and the physical referents does not have to be one-to-one if it would not normally have been so in the original cultural context of the author. Thus, no interpreter would have taken the words "God hears" and "God sees" as referring to a physical reality because such language was a clear example of accommodated speech, although their figurative meaning would still be taken as part of the *sensus litteralis* since the figurative character of the language would also have been evident to both the author and the original audience.

The church Fathers, for example, drew on accommodation in explaining why Scripture spoke of God in anthropomorphic language. Yet between the ancients and moderns stood the medieval theories of interpretation that were shared by both Jewish and Christian exegetes. Funkenstein has sketched the use of this concept among Jewish interpreters in connection with natural inquiry and suggested two different approaches to the traditional formula "Scripture speaks in human terms" (*Scriptura humane loquitur*), the maximalist and the minimalist.[8] The former, practiced by the mainstream Jewish exegetes in medieval Europe (e.g., Sa'adia, Ramban, Sforno), tended to see the biblical texts as containing full-blown scientific information couched in metaphorical language. The exegete's task consisted of a kind of translation from biblical language to philosophical truth, the result being a clear demonstration of the scientific validity of the Bible. The minimalist-contextualist approach was practiced by Ibn Ezra, who probably directly influenced Calvin and others who had studied Hebrew. We find Ibn Ezra advising that science should be learned from the Greeks because the Scriptures neither contradict science nor contain it. And the explanation of Genesis 1:16 (the two great lights) we observed in many early moderns (Calvin, Kepler, Brahe, Lansbergen) is already found fully developed in Ibn Ezra.[9]

Like their medieval Jewish and Christian predecessors, the early modern exegetes were fully persuaded of the authority of the Bible. Participants on both sides of the controversies over the motion of the earth embraced the Bible's divine inspiration. Tycho believed the Bible to be divine writ no less than Vallés his opponent, Kepler the astronomer no less than Haffenreffer the theologian. However, Funkenstein's taxonomy offers categories to understand the nature of their differences. Kepler's lengthy discussion of accommodation in the introduction to the *Astronomia Nova* follows Ibn Ezra's minimalist approach not only in downplaying the scientific content of Scripture but in its attempt to explain the language of the text by appealing to other contexts, both near and far. His use of Psalm 104 and the structure of the Genesis creation-narrative itself provides him with internal, contextual reasons to think that the divine author of the Bible did not intend the kind of genre Kepler knew as *physica*. The same procedure appears in Tycho's and Peucer's exchange over the nature of celestial matter. Both assumed and acted on the belief that the proper meaning of certain biblical phrases was best ascertained by comparing other texts with the one in question. This contextual comparison was common practice among interpreters, and it shows how attempts to resolve inconsistencies were not simply pitting a specific text against a scientific theory. The meaning of the text was examined by the expected rules of exegesis, even if the disputants could not agree on the application of the rules.

The differences between Tycho and his correspondents show that simply recognizing the presence of accommodated language in the Bible was insufficient to resolve interpretative problems. There were no interpretative difficulties for someone who adopted Rothmann's view, because for him there was no scientific content in the Bible at all, a kind of extreme minimalism in the tradition of Ibn Ezra. Extreme caution is in order with regard to Rothmann because of the complexities of the textual evidence. In his correspondence with Tycho he referred to a work on astronomy that he began in Wittenberg but that was never published. The only major study of this manuscript has shown it to be a collage of different systems, probably notes that Rothmann kept for himself.[10] Even his correspondence published by Brahe presents a sudden and unexpected change from geocentrism to heliocentrism, a radical shift that seems to have no precedent in his earlier thought. The only solid clues we have are his statements about biblical relevance in his interchange with Tycho after he had adopted the Copernican system.[11] It may be that Rothmann had become so frustrated in attempting to convert Copernican parameters to a geostatic model that he finally yielded to heliocentrism in desperation. This shift required him to rely on mathematics completely and dismiss any relevance of the biblical text at all. Rothmann would then not have to interpret specific texts because the divine author had intended no science behind the theological language of the Bible.

Both Tycho and Peucer also recognized the presence of accommodated language in the Bible. As I argued in chapter 3, Peucer stood closer to what Funkenstein called the maximalist approach, a view in which the specific biblical language referred to physical entities that might not be accessible to observation. Tycho held to an intermediate position that took the language of Genesis as adjusted to human observers but also saw that language as a possible window on cosmological truth. Tycho saw *both the Bible and cosmology* as too important to human life and morality to allow the former to be dismissed outright in the pursuit of the latter. Tycho emerges as a transitional or intermediate figure in three respects: his geoheliocentric system, his skeptical realism, and his hermeneutical moderation.

Copernicans like Rheticus, Kepler, and Lansbergen had a much more subtle approach to the Bible. They argued, of course, that the Bible did not contain scientific content like an astronomical or physical text but, at the same time, they all believed that the truths taught in the Bible were related to and embodied in the universe. Rheticus, as we observed, invoked the Bible against certain conclusions of natural philosophy (e.g., Aristotle) and was even willing to admit oblique references to the earth's motion in the Bible, an approach that was closer to Tycho's than to Rothmann's. My analysis of Kepler in chapter 4 suggests that viewing his hermeneutics in the same vein as Galileo's or Foscarini's distorts the complexity of his thought and confuses levels of analysis that are required to make sense of his uses of the Bible.[12] Lansbergen appears to have made the most vigorous argument for the separation of natural philosophy and biblical interpretation, but even he invoked biblical texts and concepts to buttress the theological desirability of a Copernican cosmology. These Copernicans appealed to accommodation as *one* of their hermeneutical strategies, but their entire repertoire of interpretation far exceeded this one feature. The subtlety of their approach sought the underlying theological truths of a text that were necessary to view nature in a Christian manner. For them, theology would never limit the possible conclusions of science, but it would certainly guide their personal reflections.

This venerable company of astronomers, philosophers, and theologians created no new exegetical strategies; they simply drew on older traditions and extended their application to a new (or revived) problem. The subtle and varied gradations in hermeneutical method reflected the historical breadth of the *sensus litteralis* so that accommodation still lay within its scope. When these disputants favored or countered appeals to literal interpretation, they were in fact implying degrees of literalness, not rigid categories. The exegetical principles various interpreters employed were not radically different from one another, but they did differ in their judgments about the divine intention in the human composition of Scripture. Those judgments were demanded by the process of understanding a written text and the special onus associated with their belief in the divine origin of Scripture. What they inferred from various texts was guided by their prior knowledge of both Christian theology and natural philosophy.

By the end of the seventeenth century, Europe looked very different scientifically and religiously than it did in 1500. At the advent of the sixteenth century, Christian Europe professed a common faith rooted in the biblical and patristic witnesses, even if the currents of deep division were brewing below the surface. At the turn of the eighteenth century, Europe resembled a patchwork quilt of religious creeds, practices, liturgies, and polities. The intervening two centuries brought such sweeping changes in the religious lives and beliefs of Europe's inhabitants that it was impossible to speak any longer of a Christian Europe as a unitary phenomenon. Those same two centuries also saw such a thoroughgoing transformation in the views of nature that the dominant Aristotelian framework of 1500 had all but given way to a collection of natural philosophies as far from Aristotle himself as Protestantism was from Catholicism. The sun-centered universe, so widely accepted among initiates, acted as a catalyst for multifaceted changes in other arenas of natural inquiry.

The divisions of modern Europe that shaped the Catholic-Protestant divide, however, did not correspond to the divisions between adherents to the old and new orders in the sciences. As we have seen, the Lutherans and Reformed were as divided among themselves over the new cosmologies as they were from Catholics over strictly theological issues. Attempts to forge strong links between various religious bodies and particular scientific viewpoints in the early modern period must ultimately fail for several reasons. On a superficial level, any strong claim advanced in favor of a Protestant ethos of science in the vein of Merton or Hooykaas has the obvious contradictions of a Galileo and a Descartes, a clerical Mersenne and a converted Nicholas Steno. On a deeper plane, the Lutherans and the Reformed differed from one another not only in strictly theological matters but in the spirit with which they approached the interface of natural philosophy and scriptural interpretation. Historians of theology have long recognized subtle but profound differences between Lutheran confessionalism and Reformed rationalism, a difference that has emerged again in this study. One effect of this difference was the greater degree to which Reformed thinkers in the Netherlands were ready to subject the Bible to individual readings. The Reformed creeds did not apparently have the same centripedal force as their counterparts in Lutheran lands.

The hermeneutical strategies employed in these cosmological debates also crossed the Catholic-Protestant divide. The historical record indicates that there were different hermeneutical traditions operating within Protestantism that had counterparts in Catholic writers. The two primary approaches discussed—*sensus litteralis* as physical description and as historical context—were employed by both Protestants and Catholics in their attempt to resolve natural knowledge with theological commitments. Yet the sixteenth and seventeenth centuries presented a new challenge due to the new cosmologies and to an emerging emphasis on the historical meaning of the biblical text. It is enticing to think that the emphasis on literal interpretation was a reaction provoked directly by the Copernican astronomy, but that would be giving too much historical weight to the new astronomy. Rather, the stress on literal interpretation in the sixteenth century derives from an emerging historical consciousness in the modern era and the dynamics of Protestant-Catholic acrimony. This dual background in fact explains partially why literal interpretation became such an important constituent in the cosmological debates of the early modern period. Literal interpretation was certainly *a part* of the historical dynamics, but it cannot be seen as cause of the emergence of the new sciences.

Contemporary reflection on science and religion requires constant attention to the historical sources of their interactions. Such reflections falter to the extent that they rely on incomplete and inaccurate historical narratives. This study, conjoined with other richly textured accounts that have emerged recently, weaves a tapestry of relations that signal stimulus, mutual support, conflict, and cooperation. Although historians are reluctant to pronounce on the essential relations between science and religion, our historical knowledge can act as a boundary condition to fend off unreliable views of those relations. One such view envisions an inherent conflict between science and religion because they both attempt to offer theories about the same field of knowledge. This view, espoused so vehemently by Andrew White and his intellectual posterity, falters on the ground of history because knowledgeable historical participants usually respected the methods and internal logic of each discipline. They rarely tried to resolve issues of conflict by imposing one discipline upon another. Another extreme view sees science and religion on two sides of a dividing wall that can never be bridged because they treat two separate aspects of human experience. This view, often developed in reaction to the former one, also fails to account for the deep desire of historical figures to obtain a unified view of knowledge in which nature and morality are seen as complementary. History may never decide the complex question of the proper relations of science and religion, but it must be a criterion which each putative theory meets.

Notes

Introduction

1. Galileo Galilei, *Il Saggiatore*, in *Opere*, vol. 6, p. 232, translated in Stillman Drake, *Discoveries and Opinions of Galileo*, p. 238.

2. John Henry Draper, *A History of the Conflict Between Religion and Science* (New York, 1874), and Andrew Dickson White, *History of the Warfare of Science with Theology in Christendom* (New York: Appleton, 1896). White organized his two-volume work according to each science in which "the sacred theory" was replaced by the true one. Chapter 3 treated astronomy. Perhaps the most thorough discussion and critique of the warfare historiography is James R. Moore, *The Post-Darwinian Controversies: A Study of the Protestant Struggle to Come to Terms with Darwin in Great Britain and America, 1870–1900* (New York: Cambridge University Press, 1979).

3. Draper and White wrote at a time when the memory of conflicts between Darwinism and Christian theology were fresh. For a possible explanation of the cultural forces influencing their histories, see John H. Brooke, *Science and Religion: Some Historical Perspectives* (Cambridge: Cambridge University Press, 1991), pp. 34, 35, and David C. Lindberg and Ronald L. Numbers, "Beyond War and Peace: A Reappraisal of the Encounter between Christianity and Science," reprinted in J. I. Packer, ed., *The Best in Theology*, vol. 1, (Carol Stream, Ill.: Christianity Today, 1988), pp. 133–49.

4. White cites Luther's *Table Talks* and Melanchthon's *Initia Doctrinae Physicae*. See *History of the Warfare*, pp. 126, 127.

5. White quoted Calvin's supposed comment on Psalm 93. See Edward Rosen, "Calvin's Attitude toward Copernicus," *Journal of the History of Ideas* 21

(1960): 431–41, for an explanation of how Calvin's famous quotation of Psalm 93 came to be wrongly attributed to him.

6. See White, *History of the Warfare*, p. 129.

7. Lindberg and Numbers explain how White shifted from discussing an undifferentiated "religion" in his earlier works to the distinction between religion and theology in his 1896 *History of the Warfare*; see David C. Lindberg and Ronald L. Numbers, "Beyond War and Peace," p. 147. White also projected the struggles of science and religion in his own experience back into the past. He "read the past through the battle-scarred glasses" of his own conflicts, Lindberg and Numbers, "Beyond War and Peace," p. 134. White made repeated references to contemporary clerics who attempted to control science research and teaching in the universities. White, *History of the Warfare*, 1:126, 128.

8. Alexandre Koyré, *The Astronomical Revolution* (New York: Dover, 1992), p. 72. See pp. 74–75 for further elaboration of biblical opposition.

9. The diversity of responses to the new sciences has been elucidated also by the recognition of a diversity of types of Aristotelianism in the Renaissance. See Edward Grant, *In Defense of the Earth's Centrality and Immobility: Scholastic Reaction to Copernicanism in the Seventeenth Century* (Philadelphia: American Philosophical Society, 1984), and Dominick Iorio, *The Aristotelians in Renaissance Italy: A Philosophical Exposition* (Lewiston, N.Y.: Edwin Mellen Press, 1991). In one of the most comprehensive surveys of science and religion since White's 1896 book, John Hedley Brooke has written what is in effect an extended example of the diversity of such interactions. For examples of the relations, see Brooke, *Science and Religion*, pp. 19–33. Brooke has offered other incisive criticisms of the conflict thesis throughout his volume but see particularly pp. 33–42.

10. See, e.g., Giorgio de Santillana, *The Crime of Galileo* (Chicago: University of Chicago Press, 1955); Ernan McMullin, ed., *Galileo, Man of Science* (New York: Basic Books, 1967), especially the editor's introduction; James J. Langford, *Galileo, Science and the Church* (Ann Arbor: University of Michigan Press, 1971) for a moderate defense of the church. This interest in the Galileo affair remains unabated today, and the literature is so vast that I can cite only several prominent works. See Pietro Redondi, *Galileo Heretic* (Princeton: Princeton University Press, 1987); Mario Biagioli, *Galileo, Courtier* (Chicago: University of Chicago Press, 1993); Annibale Fantoli, *Galileo for Copernicanism and for the Church* (Vatican: Vatican Observatory Foundation, 1994).

11. Olaf Pedersen, "Galileo and the Council of Trent: The Galileo Affair Revisited," *Journal of the History of Astronomy* 14 (1983): 1–29.

12. Richard Blackwell, *Galileo, Bellarmine, and the Bible* (Notre Dame: University of Notre Dame Press, 1991), see chap. 7.

13. Introduction, *Astronomia Nova* (1609), in Max Caspar, ed., *Keplers Gesammelte Werke*, vol. 3, p. 29.

14. John Wilkins, *The Discovery of a New World* (London, 1638).

15. Thomas Kuhn, *The Copernican Revolution* (New York: Random House, 1959), pp. 191ff.

16. Blackwell, *Galileo, Bellarmine, and the Bible*, pp. 169–71.

17. See Kenneth J. Howell, "Copernicanism and the Bible in Early Modern Science," in Jitse van der Meer, ed., *Facets of Faith and Science* (University Press of America, 1994), for a survey of the use of the Bible in the Copernican debates.

18. Peter Harrison, *The Bible, Protestantism, and the Rise of Natural Science* (Cambridge: Cambridge University Press, 1998) p. 4.

19. Ibid., p. 4.

20. Ibid., p. 4.

21. Ibid., p. 268.

22. Ibid., p. 267.

23. Ibid., p. 271.

24. Most of Harrison's sources appear to be limited to early English works or translations into English during the early modern period.

25. Ibid., p. 271.

26. Jerzy Dobrzycki, T*he Reception of Copernicus' Heliocentric Theory* (Dordrecht: D. Reidel, 1972), discusses the reception of Copernicanism in various national contexts. Newer studies precipitated something of a crisis that resulted in a loss of consensus about the term "scientific revolution." See David C. Lindberg and Robert S. Westman, Introduction, in Lindberg and Westman, eds., *Reappraisals of the Scientific Revolution* (Cambridge: Cambridge University Press, 1990). This reevaluation was spurred by attention to alchemy, among other things. On Newton, see Betty Jo Dobbs, *The Hunting of the Green Lion: The Foundations of Newton's Alchemy* (Cambridge: Cambridge University Press, 1975), and *The Janus Face of Genius* (Cambridge: Cambridge University Press, 1991). On Boyle, see Lawrence Principe, "Boyle's Alchemical Pursuits," and William R. Newman, "Boyle's Debt to Corpuscular Alchemy," in Michael Hunter, ed., *Robert Boyle Reconsidered* (Cambridge: Cambridge University Press, 1994). The emphasis on social context, of which patronage has been a part, is found in Richard S. Westfall, "Scientific Patronage: Galileo and the Telescope," *Isis* 76 (1985): 11–30. It had long been thought that universities were a conservative force which resisted the advances of science. For a more positive assessment, see John Gascoigne, "A Reappraisal of the Role of Universities in the Scientific Revolution," in Lindberg and Westman, *Reappraisals*, pp. 207–60. A more radical tack posits local frameworks of meaning

to explain how the results of science were themselves products of social construction. See Steven Shapin and Simon Schaffer, *The Leviathan and the Air-Pump: Hobbes, Boyle, and the Experimental Life* (Princeton: Princeton University Press, 1985), and Steven Shapin, *A Social History of Truth, Civility, and Science in Seventeenth-Century England* (Chicago: University of Chicago Press, 1994).

27. I. Bernard Cohen, *Revolution in Science* (Cambridge: Harvard University Press, 1985), pp. 106, 125.

28. Neugebauer's research exploded the notion that Copernicus's system was simpler than Ptolemy's. See Otto Neugebauer, "The Equivalence of Ptolemaic and Copernican Astronomy," in *Vistas in Astronomy* (1968). See also Cohen, *Revolution in Science*, p. 125.

29. The language of "monster" and "man" comes from Copernicus's Preface to *De Revolutionibus*. See Owen Gingerich, " 'Crisis' versus Aesthetic in the Copernican Revolution," in *Vistas in Astronomy* 17 (1975): 85–95, reprinted in Owen Gingerich, *The Eye of Heaven: Ptolemy, Copernicus, Kepler*, A. Beer and K. Strand, eds. (New York: American Institute of Physics, 1993), pp. 193–204.

30. Owen Gingerich, "The Great Copernicus Chase," *American Scholar* 49 (1979): 81–88, reprinted in Gingerich, *The Eye of Heaven*, p. 75.

31. Most Melanchthon studies naturally treat his theology without much regard to his natural philosophy. For studies of Melanchthon's influence on science, see note 5 in chapter 2.

32. See Gingerich, *The Eye of Heaven*, pp. 234ff.; Robert S. Westman, "The Melanchthon Circle, Rheticus and the Wittenberg Interpretation of the Copernican Theory," *Isis* 66 (1975): 165–93; "Humanism and Scientific Roles in the 16th Century," in R. Schmitz and F. Krafft, eds., *Humanismus und Naturwissenschaften* (Boppard am Rhein: Boldt, 1980), pp. 83–99.

33. Still the most extensive treatment of the problem of celestial matter in this period is William H. Donahue, *The Dissolution of the Celestial Spheres* (New York: Arno Press, 1981).

1. Reading the Heavens and Scripture in Early Modern Science

1. "[mundus] qui, propter nos, ab optimo et regularissimo omnium opifice," Nicholas Copernicus, Preface to *De Revolutionibus*, edited by Heribert Maria Nobis and Bernhard Sticker (Hildesheim: Gerstenberg Verlag, 1984), p. 4.

2. John Ray, *The Wisdom of God Manifested in the Works of Creation* (1691) (reprinted, Hildesheim: Georg Olms Verlag, 1974).

3. Thomas Sprat, *The History of the Royal Society*, quoted in John H. Brooke, *Science and Religion* (Cambridge: Cambridge University Press, 1991), p. 82. The religious utility of science came to light especially in Merton's seminal study of seventeenth-century England. Robert K. Merton, "Science, Technology and Society in Seventeenth Century England," *Osiris* 38 (1937): 360–632. Merton suggested an intimate link between Puritanism and the emergence of science. The widely discussed Merton thesis has been subject to a variety of interpretations and extensions, but Merton's originally modest claim was simply that the English Puritan religious ethos had created a climate in which the pursuit of the knowledge of nature was considered a religious duty. His work showed a disproportionally higher number of Puritans than non-Puritans in the Royal Society in the course of seventeenth-century English science. This thesis implied two important claims: that there was a positive influence of religion on science in the crucial period of early modern Europe, and more particularly that Puritanism represented a specific constellation of religious values that sanctioned science and provided a religious motivation for that activity. Merton's thesis was anticipated in the nineteenth century by Candolle, who provided statistics of a disproportionally higher number of Protestants in European science. See Alphonse de Candolle, "The Influence of Religion on the Development of the Sciences," in *Histoire des sciences et des savants depuis deux siècles; précédéé et suivie d'autres études sur les sujets scientifiques, en particulier sur l'hérédité et la selection dans l'espèce humaine*, 2nd ed. (Geneva: H. Georg, 1885), pp. 328–36, translated in I. Bernard Cohen, ed., *Puritanism and the Rise of Modern Science* (New Brunswick: Rutgers University Press, 1990), pp. 145–50. Pelseneer developed this claim further by offering statistical information on sixteenth-century scientific works from which he argued that a living faith was necessary for the establishment of any science. For him, there was a direct causal link between modern science and the Reformation. See Jean Pelseneer, "The Protestant Origin of Modern Science," translated in Cohen, *Puritanism and the Rise of Modern Science*, pp. 178–80; appeared originally as "L'origine protestante de la science moderne," *Lychnos* (1946–47): 246–48. Hooykaas and Mason buttressed this argument with historical analyses that claimed a special connection between science and the Calvinist view of life. See Reijer Hooykaas, "Science and Reformation," *Journal of World History* 3 (1956): 109–39, and Stephen F. Mason, "The Scientific Revolution and the Reformation," *Annals of Science* 9 (1953): 64–175, and, in slightly altered form, *A History of the Sciences* (New York: Collier Books, 1962), pp. 175–91. These are reprinted in Cohen, *Puritanism and the Rise of Modern Science.*

4. *De Sapientia Veterum*, The Wisdom of the Ancients, 1609, in *Francis Bacon: A Selection of His Works*, edited by Sidney Widhaft (New York: The Odyssey Press, 1965), p. 284.

5. Psalm 19:1, "The heavens declare the glory of God and the firmament showeth his handiwork."

6. Francis Bacon, *The Great Instauration* (part I of the *Novum Organum*, 1620), p. 23.

7. For a discussion of the differences between technical astronomy and natural philosophy, see Edward Grant, *Planets, Orbs and Stars: The Medieval Cosmos, 1200–1687* (Cambridge: Cambridge University Press, 1994), pp. 36–39.

8. One such astronomer was the Jesuit Christopher Clavius, whose commentary on Sacrobosco's *Sphere* repeatedly argued for astronomy as a true science. See Christopher Clavius, *In Sphaeram Iohannis de Sacrobosco Commentarius* (Rome, 1581). On Clavius, see William A. Wallace, "Galileo and Reasoning Ex Suppositione," in *Prelude to Galileo: Essays on Medieval and Sixteenth Century Sources of Galileo's Thought* (Dordrecht, 1981), pp. 129–59, and Rivka Feldhay, "Knowledge and Salvation in Jesuit Culture," *Science in Context* 1 (1987): 195–213, and more recently, James M. Lattis, *Between Copernicus and Galileo: Christoph Clavius and the Collapse of Ptolemaic Astronomy* (Chicago: University of Chicago Press, 1994).

9. Among the editions of Ptolemy's *Geography* can be found Sebastian Münster's (Basel, 1540), Mattioli's (Venice, 1548), and Giovanni Magini's (Venice, 1596). Erasmus printed a critical Greek edition (Basel, 1533). For a good discussion, see H. N. Stevens, *Ptolemy's Geography: A Brief Account of the Printed Editions Down to 1730* (London, 1908). Authors also offered their own cosmographies. In addition to the works discussed below, some of these included Pomponius Mela, *Cosmographia sive de situ orbis* (late fifteenth century), and Franciscus Barocius, *Cosmographia* (Venice, 1585).

10. Francesco Maurolico, *Cosmographia* (Venice, 1543), "qui nunquam à speculatione, nunquam à contemplatione vacat," p. 1. On the necessity of reason (*opus est ratione*), see p. 5.

11. "Cosmography, as its etymology makes clear, is a description of the world that consists of the four elements (earth, water, air, fire) as well as of the sun, moon and all other stars, in fact whatever is under the revolving of the heavens," Petrus Apianus, *Cosmographia* (Antwerp, 1584), p. 1. The 1584 edition was supplemented with a work by Gemma Frisius, *De Radio Astronomico*.

12. "It differs from geography because it treats the earth only by referring to the circles of the sky, not by reference to mountains, seas, and rivers, etc." Petrus Apianus, *Cosmographia* (Antwerp, 1584), p. 1.

13. "It is fitting furthermore that you be congratulated with many new orbs that are now known for the first time," Girolamo Fracastoro, *Homocentria eiusdem de causis criticorum dierum per ea quae in nobis sunt*, Preface, unpaginated (Venice, 1538).

14. "In our homocentrics there is not only the utility that follows all astronomy but also other things that lead first of all to truth itself," Fracastoro, *Homocentria*, chapter 1, on astronomy. Fracastoro was only representative of a larger Paduan philosophy that attempted a revival of the Eudoxian spheres, a leading proponent of which was Giovan Battista Amico. See Mario Di Bono, *Le Sfere Omocentriche di Giovan Battista Amico Nell'Astronomia del Cinquecento*, con il testo *De motibus corporum coelestium* (Genova: Consiglio Nazionale delle Richerche, 1990).

15. For a general description of Croll's work, see Allen G. Debus, *The Chemical Philosophy: Paracelsian Science and Medicine in the Sixteenth and Seventeenth Centuries* (New York: Science History Publications, 1977), vol. 1, pp. 117ff.

16. Oswald Croll, *Basilica Chymica* (1609), 1st ed. Eng. trans. by Rev. Pennde in 1657.

17. Croll's Calvinism differed from Calvin's own theology in significant ways. Croll's transformation of Paracelsus placed its religious aspects in a new theological context but not without alteration of the context itself. For an exposition of Calvin and Croll, see Owen Hannaway, *The Chemists and the Word: The Didactic Origins of Chemistry* (Baltimore: The Johns Hopkins University Press, 1975), pp. 52–57. For a more recent exposition of Croll's linking of nature and the word of God, see James J. Bono, *The Word of God and the Languages of Man* (Madison: University of Wisconsin Press, 1995).

18. Croll, *Basilica Chymica* (1609), p. 48.

19. See Debus, *The Chemical Philosophy*, pp. 160, 180.

20. Croll, *Basilica Chymica* (1609), p. 48.

21. William Derham, *Astro-theology* (London: W. Innys, 1715), Preface, p. xl.

22. On Derham's belief that other inhabitable worlds existed, see ibid., xliv–lvii.

23. End of the Preface, ibid., lviii.

24. "Which language of the heavens is so plain and their characters so legible that all, even the most barbarous nations, that have no skill either in languages or letters, are able to understand and read what they proclaim," Ibid., book 1, p. 2.

25. Ibid., chapter 3, p. 17.

26. Ibid., chapter 3, p. 57.

27. Ibid., book 4, chapter 1, pp. 61–67.

28. For a broad survey of the natural theology tradition in England, see John Hedley Brooke, *Science and Religion: Some Historical Perspectives* (Cambridge University Press, 1991), chapter 6, pp. 192–225. This tradition was transplanted to New England and was most thoroughly expressed by Cotton Mather. His *Christian Philosopher: A Collection of the Best Discoveries in Nature with Religious Improvements* (London, 1721) extended this detailed comparison of the two books of nature and Scripture by surveying the great discoveries of the previous century and drawing religious morals from them. See Winton U. Solberg, "Science and Religion in Early America: Cotton Mather's *Christian Philosopher*," *Church History* 56 (1987): 73–92.

29. *De errore sectariorum huius temporis Labyrintheo*, pp. 211–12.

30. We are largely ignorant of this hexaemeral tradition, and our knowledge of Renaissance commentary on Genesis is quite dated and incomplete. Still today the only book-length treatment of Renaissance hexaemeral commentaries is Arnold Williams, *The Common Expositor: An Account of the Commentaries on Genesis, 1527–1633* (Chapel Hill: University of North Carolina Press, 1948). Williams's book had a literary focus and treated scientific issues with respect to Genesis only in one chapter (9), pp. 174–98.

31. Minucius Felix, *Octavius*; selected passages reprinted in Peter E. Herbert, *Selections from the Latin Fathers* (Boston: Ginn, 1924), pp. 11ff. A general description of this dialogue can be found in Johannes Quasten, *Patrology* (Westminster, Md.: Christian Classics), first published 1950, sixth reprinting in 1992, vol. 2, pp. 155ff. See Herbert's notes on p. 131 for a comparison between Municius and Cicero in their wording: "How can it be so obvious, so incontrovertible, and so clear than that there is some deity with the most exceptional mind when you lift your eyes to heaven and you survey what is below and above?"

32. Herbert, *Selections from the Latin Fathers*, "Summus Moderator," p. 11; for Municius's detailed descriptions, see p. 13.

33. The Greek text of Basil's lecture can be found in *To Students on Greek Literature*, notes and vocabulary by Edward Maloney (New York: American Book Company, 1901); see esp. p. 19.

34. "I know the laws of allegory even if not from my own works, from those of others. They do not take the everyday meanings of words. So water doesn't mean water but indicates some other nature. And so with . . . They interpret as it seems to them. . . . But when I hear grass, I think of grass. And so with. . . ." Basil the Great, *Homilies on the Six Days of Creation*. All translations are my own from a comparison of the Greek and Latin texts.

35. For comparisons between Basil and Plato's *Timaeus*, see Frank Robbins, *The Hexaemeral Literature: A Study in the Greek and Latin Commentaries on Genesis* (Ph.D. diss. at University of Chicago, 1912), pp. 42ff.

36. This word order is found in the Massoretic text, the Greek Septuagint, the Old Latin, and Vulgate translations.

37. Homily 1, *On the Hexaemeron*, sec. 1, *Patrologia Graece* 4.

38. Basil, *On the Hexaemeron*. I have translated this passage from the quotation given by Robbins, *The Hexaemeral Literature*, pp. 44, 45.

39. Basil's *Hexaemeron* had considerable influence on Ambrose of Milan, but the latter took a different approach to Genesis. His homilies on the six days were strongly oriented to the moral implications of the natural world and interacted little with the natural philosophy of the ancient world. Selections from Ambrose's *Hexaemeron* can be found in Peter E. Herbert, *Selections from the Latin Fathers* (New York: Ginn, 1924), pp. 51ff.

40. Augustine "made appeal to a cosmology and a physics (time, unformed matter), to a natural theology (motive for creation, transcendence of the Spirit's action), to a logic (concept of darkness), and even to a psychology (analogy of human will and divine will)," Aimé Solignac, "Exégèse et Métaphysique Genèse 1:1–3, chez saint Augustin," in *In Principio Interpretations des premiers versets de la Genese, Etudes Augustiniennes* (Paris: C.N.R.S. 1973), pp. 153–71; quotation on p. 157.

41. Augustine attempted to comment on creation narratives of Genesis on five separate occasions during his life. His earliest but aborted attempts are translated in *On the Literal Interpretation of Genesis: An Unfinished Book* and *Two Books on Genesis against the Manichees*, trans. Roland J. Teske (Washington, D.C.: Catholic University Press, 1980). Then he explored the same texts in his *Confessions*. See the recent translation by Owen Chadwick, *Confessions* (Oxford: Oxford University Press, 1991). His longest commentary appeared in *De Genesi ad Litteram* which has been translated by John Hammond Taylor as *The Literal Meaning of Genesis* (New York: Newman Press, 1982), 2 vols. His final attempt was *De Civitate Dei*. In all his exegesis, Augustine never proposed any definitive solutions to the interpretative problems he posed for himself.

42. Augustine held that the creation of the world in an instant was known to the angels because they, unlike humans, had an immediate knowledge of eternal forms. Because the angels are beyond time, they can perceive that which humans can understand only through a sequence of events. Humans need a temporal description to understand a nontemporal reality. For an exposition of Augustine's doctrine of creation, see William A. Christian, "Augustine on the Creation of the World," *Harvard Theological Review* 46 (1953): 1–25.

43. For Augustine's polemic against the Manichees with respect to Genesis, see John P. Maher, "Defense of the Hexaemeron," *Catholic Biblical Quarterly* (part III = vol. 7 [1945], pp. 76–90).

44. The medieval period witnessed a long stream of hexaemeral commentary. Especially important for the early modern period were the commentaries of Thomas Aquinas in his *Summa Theologica* (Qq. 65–74), Bonaventure in his *Hexaemeron*, and Henry of Langenstein. On Langenstein, see Nicholas H. Steneck, *Science and Creation in the Middle Ages* (Notre Dame: University of Notre Dame Press, 1976).

45. For a general overview of Luther's hermeneutics, see Willem Jan Kooiman, *Luther and the Bible* (Philadelphia: Muhlenberg Press, 1961); Heinrich Bornkamm, *Luther and the Old Testament* (Philadephia: Fortress Press, 1969); James Samuel Preus, *From Shadow to Promise: Old Testament Interpretation from Augustine to the Young Luther* (Cambridge: Harvard University Press, 1969); David C. Steinmetz, *Luther and Staupitz: An Essay in the Intellectual Origins of the Protestant Reformation* (Durham: Duke University Press, 1980).

46. Martin Luther, *Lectures on Genesis*, chap. 1–5 in *Luther's Works*, vol. 1, ed. Jaroslav Pelikan. Translated by George Schick (St. Louis: Concordia Publishing House, 1958). I here rely on Schick's standard translation.

47. Ibid., pp. 3, 4. Luther also speaks of his task as a "journey without a guide" (p. 6). See also his perjorative comments on philosophy: "let us turn to Moses as the better teacher. We can follow him with greater safety than the philosophers" (p. 6).

48. For further explanation of this tendency, see Peter Harrison, *The Bible, Protestantism, and the Rise of Natural Science* (Cambridge: Cambridge University Press, 1998).

49. Among the extensive literature on Calvin's hermeneutics, see Gottfried W. Locher, *Testimonium Internum Calvins Lehre vom Heiligen Geist and das hermeneutische Problem* (Zurich: evangelischer Verlag, 1964); Alexandre Ganoczy and Stepfan Scheld, *Die Hermeneutik Calvins Geistesgeschichtliche Voraussetzung und Grundzüge* (Wiesbaden: Franz Steiner Verlag, 1983); T. H. L. Parker, *Calvin's Old Testament Commentaries* (Edinburgh: T. & T. Clark, 1986); Susan Scheiner, *Where Shall Wisdom Be Found? Calvin's Exegesis of Job from Medieval and Modern Perspectives* (Chicago: University of Chicago Press, 1994); Wilhelm H. Neuser, *Calvinus Sacrae Scripturae: Professor Calvin as Confessor of Holy Scripture* (Grand Rapids: Eerdmans, 1994).

50. In Book 1, chapter 1, Augustine argues that taking "the facts that are narrated" in a historical sense is justified. Taking the narrative in a figurative sense is unproblematic. See *The Literal Meaning of Genesis* (Taylor, ed.), p. 19.

51. This is a relatively unexplored area. One of the most important contributions to understanding the history of hermeneutics is Henri de Lubac, *Exégèse Mediévale: Les Quatre Sens de l'Ecriture* (Paris: Aubier, 1959), 4 vols.

52. Zanchius, prefatory letter of his *De operibus Dei intra Spacium Sex Dierum Creatis* (original, 1591), reprinted in *Operum Theologicorum*, vol. 3 (1619).

53. Ibid.

54. Arnold Williams, *The Common Expositor*, p. 178.

55. A general survey of Mersenne's life and work can be found in Peter Dear, *Mersenne and the Learning of the Schools* (Cornell: Cornell University Press, 1988).

56. The whole title is *Quaestiones celeberrime in Genesim, cum accurata textus explicatione* (Paris, 1623).

57. "In hoc volumine athei, deistae impugnantur, et expurgantur, et Vulgata editio ab haereticorum calumniis vindicantur," Mersenne, *Quaestiones* preface.

58. "Graecorum et Hebraeorum musica instauratur. . . . Opus theologis, philosophis, medicis, iuriconsultis, mathematicis, musicis vero, catoptricis praesertim utile," Mersenne, *Quaestiones* preface.

59. Still the best available articles analyzing Mersenne's Genesis commentary are William L. Hines, "Mersenne and Copernicanism," *Isis* 64 (1973): 18–32, and idem, "Marin Mersenne: Renaissance Naturalism and Renaissance Magic," in Brian Vickers, *Occult and Scientific Mentalities in the Renaissance* (Cambridge: Cambridge University Press, 1984), pp. 165–76. A general description of Mersenne's exegetical work can be found in Albano Biondi, "L'esegesi biblica di frate Marin Mersenne," *Annali di storia dell'esegesi* 9 (1992): 35–52.

60. *Quaestiones*, cols. 1147–48.

2. Copernicus, the Bible, and the Wittenberg Orbit

1. Copernicus identifies Aristarchus as the originator of heliocentricism, but very quickly astronomers in the sixteenth century began to speak of the new theory as Copernican. Our knowledge of Aristarchus of Samos largely depends on other ancient writers. Archimedes described his heliocentric view with censure in his *Psammites*. Simplicius reported Aristarchus's view in his quotation from Geminus in his *Physica*. For translations of these texts see Thomas L. Heath, *Greek Astronomy* (London: J. M. Dent, 1932; reprinted by Dover, 1991).

2. Tycho Brahe, for one, admired Copernicus's systemic superiority to Ptolemy. See the discussion in chapter 3. Historians have followed Copernicus's suggestions that the "monster" of Ptolemy had to be simplified. See Thomas S. Kuhn, *The Copernican Revolution* (New York: Random House, 1959) p. 76.

3. The historical value of the *Table Talks* has been examined in Heinrich Bornkamm, "Kopernikus im Urteil der Reformation," *Archiv für Reformationsgeschichte* 40 (1973). See the discussion in Hans Blumenberg, *The Genesis of the Copernican World* (Cambridge: MIT Press, 1987), p. 320. Even if we accept Luther's remark ("The fool [Copernicus] will attempt to overturn the whole art of astronomy") as historically accurate, it is not clear that Luther was opposed to the *content* of Copernicus's theory as such.

4. Westman calls attention to White's and Draper's contention that Luther's words negatively influenced the reception of Copernicus in Protestant areas of Europe. Robert S. Westman, "Copernicus and the Churches," in David Lindberg and Ronald Numbers, eds., *God and Nature: Historical Essays on the Encounter between Christianity and Science* (Berkeley: University of California Press, 1986), p. 82. See also Blumenberg's discussion of different interpretations of Luther's remarks. Blumenberg, *The Genesis of the Copernican World* (1987), pp. 320, 321. I cannot, however, endorse Blumenberg's view of Luther's supposed "fearfulness of the heavens."

5. Among the extensive literature on Melanchthon (and other Wittenberg figures) the relevant material on natural philosophy includes J. R. Christianson, "Copernicus and the Lutherans," *Sixteenth Century Journal* 4 (October 1973): 1–10; Bruce T. Moran, "The Universe of Philipp Melanchthon: Criticism and Use of the Copernican Theory," *Comitatus* 4 (1973): 1–23; Zofia Wardeska, "Copernicus und die deutschen Theologen des 16. Jahrhunderts," in F. Kaulbach, *Nicholaus Copernicus zum 500 Geburtstag* (Cologne: Böhlau Verlag, 1973), pp. 155–84; Robert S. Westman, "The Melanchthon Circle, Rheticus, and the Wittenberg Interpretation of the Copernican Theory," *Isis* 66 (1975): 165–93; idem, "The Astronomer's Role in the Sixteenth Century: A Preliminary Study," *History of Science* 23 (1980): 105–47. The latest comprehensive study of Melanchthon's natural philosophy can be found in Sachiko Kusukawa, *The Transformation of Natural Philosophy* (Cambridge: Cambridge University Press, 1995).

6. For the sociopolitical context of Lutheran confessionalism, see Heinz Schilling, "Die Konfessionalisierung im Reich, Religiöser und gesellschaftlicher Wandel in Deutschland zwischen 1555 und 1620," *Historische Zeitschrift* 246 (1988): 1–45 and idem, "Reformation und Konfessionalisierung in Deutschland und die neuere deutsche Geschichte," *Gegenwartskunde, Gesellschaft, Staat, Erziehung* Sonderheft 6 (1988): 11–29. See also Robert Kolb, *Confessing the Faith: Reformers Define the Church, 1530–1580* (St. Louis: Concordia, 1991).

7. Kolb, *Confessing the Faith*, p. 26.

8. Luther attempted to show contradictions between various Fathers as well as some Fathers' belief that Scripture alone was inerrant. He cites Augus-

tine's letter to Jerome as a means of showing disagreements among the Fathers: "On the Councils and the Church" (1539), *Luther's Works: Church and Ministry*, 3, ed. Eric W. Gritsch (Philadelphia: Fortress Press, 1966), vol. 41, p. 25.

9. Enthusiasm admits of varying definitions and interpretations. Most of Luther's invectives are directed against those Anabaptists who tended to rely on private revelations and deemphasized the ministry of the church. For a discussion of Enthusiasm, see L. McCann, "Religious Enthusiasm," *New Catholic Encyclopedia* (New York: McGraw-Hill, 1967), vol. 5, pp. 446–49.

10. Martin Luther, "Against the Papacy, an Institution of the Devil," in *Luther's Works: Church and Ministry*, 3, vol. 41.

11. "We believe this not on account of that bishop, but because we see that this view is taught in the Word of God, and the testimonies of the ancient church agree with it. Likewise I say that synods of the church are to be heard which teach and remind us when they discuss the Word of God. But they should be examined and when they draw right conclusions, we believe because of the Word of God" (1539), *De ecclesia et de authoritate verbi Dei*, in *Melanchthon's Werke*, vol. 1, Robert Stupperich, ed. (Gütersloh: Bertelsmann Verlag, 1951), p. 338. An English translation of this tract can be found in Ralph Keen, *A Melanchthon Reader* (New York: Peter Lang, 1988), p. 248.

12. 1539, *De ecclesia et de authoritate verbi Dei*, in *Melanchthon's Werke*, vol. 1, p. 326. English trans. in Keen, *A Melanchthon Reader* (New York: Peter Lang, 1988) p. 239.

13. The term was first used by Abraham Calov, *Syncretismus Calixtinus* (Wittenberg, 1653), pp. 318ff., and is discussed in Robert D. Preus, *The Theology of Post-Reformation Lutheranism* (St. Louis: Concordia, 1970), vol. 1, p. 38.

14. The most important controversy occurred in 1549 when Osiander, professor at Königsberg, opposed what he thought an overemphasis on the forensic nature of justification made by Luther and Melanchthon. See the entry under "Osiandrian Controversy" in Erwin Lueker, *Lutheran Cyclopedia* (St. Louis: Concordia, 1975), rev. ed., p. 593.

15. See "Osiander, Andreas" in Lueker, *Lutheran Cyclopedia*, p. 593.

16. See Pierre Duhem, *To Save the Phenomena: An Essay in the Idea of Physical Theory from Plato to Galileo* (Chicago: University of Chicago Press, 1969), and *The Aim and Structure of Physical Theory* (Chicago: University of Chicago Press, 1912). On Duhem's views of science and religion, see Stanley L. Jaki, *Uneasy Genius: The Life and Work of Pierre Duhem* (The Hague: Martinus Nijhoff, 1984), and R. N. D. Martin, *Pierre Duhem: Philosophy and History in the Work of a Believing Physicist* (La Salle, Ill.: Open Court, 1991).

17. Robert S. Westman, "The Astronomer's Role in the Sixteenth Century: A Preliminary Study," *History of Science* 23 (1980): 105–47.

18. *Osiander to Joachim Rheticus*, (Nuremberg, 1541), in Andreas Osiander, *Gesamtausgabe*, vol. 7, *Schriften and Briefe* (Gutersloher Verlaghaus), p. 337.

19. *Osiander to Copernicus* (Nuremberg, 1541), in Osiander, *Gesamtausgabe*, pp. 333–35.

20. *Ad lectorem de hypothesibus huius operis*.

21. See the comments by Heiko Oberman and Hans-Ulrich Hofman in Osiander, *Gesamtausgabe*, pp. 337–38, n. 7.

22. See Erwin Lueker, *Lutheran Cyclopedia*, under "Osiander Controversy" (St. Louis: Concordia, 1975), p. 593.

23. The text of his oration for the Nuremberg school, "In laudem novae scholae," is in Robert Stupperich, *Melanchthons Werke in Auswahl*, vol. 3 (Gütersloh: C. Bertelsmann, 1951), pp. 64–69 and is translated as "In Praise of the New School," in Ralph Keen, *A Melanchthon Reader* (New York: Peter Lang, 1988). Melanchthon's role in founding the Nuremberg school is discussed in W. Maurer, *Der Junge Melanchthon zwischen Humanismus und Reformation* (Göttingen, 1969).

24. Gerhard Müller, "Philipp Melanchthon zwischen Pädagogik und Theologie," in *Humanismus im Bildungswesen des 15. und 16. Jahrhunderts*, Wolfgang Reinhard, ed. (Weinheim: Acta Humaniora, 1984), pp. 95–106. Müller suggests that Melanchthon's *privata schola* in his Wittenberg home may have been meant to show an example of piety, since Melanchthon firmly believed in the principle "*pietas pietate discitur*"; see pp. 97–99.

25. The inaugural address entitled "De corrigendis adolescentiae studiis" is in Robert Stupperich, *Melanchthons Werke in Auswahl*, vol. 3 (Gütersloh: C. Bertelsmann, 1951), pp. 30–42 and is translated as "On Correcting the Studies of Youth," in Ralph Keen, *A Melanchthon Reader* (New York: Peter Lang, 1988), pp. 47–63.

26. On Reuchlin and his relationship with Melanchthon, see Ralph Keen, Introduction, *A Melanchthon Reader*, pp. 1–5.

27. See *Melanchton to Oecolampadius*, 8 April 1529, translated in Keen, *A Melanchthon Reader*, pp. 127–29. See also the discussion in Gordon Rupp, "Philipp Melanchthon," in Hubert Cutliffe-Jones, ed., *A History of Christian Doctrine* (Philadelphia: Fortress Press, 1978), p. 377.

28. See Westman, "The Melanchthon Circle, Rheticus, and the Wittenberg Interpretation of the Copernican Theory," *Isis* 66 (1975): 168–74 and Bruce T. Moran, "The Universe of Philipp Melanchthon," *Comitatus* 4 (1973): 5–7.

29. "Atheist" in the sixteenth century did not generally carry the connotation of a wholesale denial of divine existence but a denial of fundamental propositions which would lead to knowledge of God. See Kusukawa (1992), pp. 161, 162. The reference to Genesis 1:14 is found in *Melanchthon to Grynaeus*

(1531), "if anyone also requires some authority from the Sacred Scriptures that commends this study, he has it in the weightiest testimony: they are in signs, time and years," *Corpus Reformatorum* ii, p. 531f.

30. So argued by Sachiko Kusukawa, *The Transformation of Natural Philosophy* (Cambridge: Cambridge University Press, 1995).

31. *Loci Communes* (1543), trans. by J. A. O. Preus (St. Louis: Concordia, 1992), p. 70.

32. Melanchthon sees the effect of sin in obscuring human acknowledgment of God as due to both the fall of the original human pair and to an active suppression of this knowledge on the part of individuals. See ibid., p. 70.

33. On the composition and prehistory of the *Initia*, see Kusukawa, *The Transformation of Natural Philosophy*, p. 184.

34. Ibid., p. 169.

35. Melanchthon's actual definition of *physica doctrina* is: "that which inquires into and lays bare the regularity, qualities and motions of all bodies and species in nature along with the causes of generation and corruption as well as of other motions in the elementary [realm] and other bodies that are made up of a mixture of elementary substances to the extent that this can be understood by the darkness of the human mind" (*Initia doctrinae physicae*, p. 181).

36. Ibid., p. 179.

37. Ibid., p. 181.

38. See *Commentary on Romans*, trans. Fred Kramer (St. Louis: Concordia, 1992) pp. 179–81.

39. *Initia doctrinae physicae*, p. 192.

40. Ibid., p. 216.

41. Ibid., p. 217.

42. Ibid., p. 186.

43. Ibid., pp. 186 and 214.

44. "If the world were infinite, an infinite confusion of nature [rerum] and the arts would follow," ibid., p. 214.

45. On Melanchthon's use of the patristic sources, see Peter Fraenkel, *Testimonium Patrum: The Function of the Patristic Argument in the Theology of Philipp Melanchthon* (Geneva: Libraire E. Droz, 1961).

46. The diversity of Rheticus's interests are described in Edward Rosen, "Rheticus, George Joachim," in the *Dictionary of Scientific Biography* (Washington: American Council of Learned Societies, 1975), vol. 11, pp. 395–98. In his first teaching position in Wittenberg (1536–38), Rheticus, although responsible for arithmetic and geometry, read so extensively in astronomy (Erasmus Reinhold's domain) that he was dubbed "The Astronomy Professor." See Karl H.

Burmeister, *Georg Joachim Rheticus, 1514–1574, Eine Bio-Bibliographie*, 3 vols. (Wiesbaden: Guido Pressler Verlag, 1966), vol. 1, p. 31.

47. *Tiedemann Giese to Rheticus*, in Burmeister, *Rheticus*, vol. 3, pp. 54, 55.

48. Reijer Hooykaas, *G. J. Rheticus's Treatise on the Holy Scripture and the Motion of the Earth*, with translation, annotations, commentary, and additional chapters on Ramus-Rheticus and the development of the problem before 1650. Verhandelingen der Koninklijke Nederlandse Akademie van Wetenschappen, Afd. Letterkunde, Nieuwe Reeks, Deel 124. I will cite Rheticus's document as *Rheticus's Treatise*, followed by reference to the section number and the page number of Hooykaas's English translation. I will refer to Hooykaas's *Commentary* by the page number of this work.

49. See Hooykaas's comparisons, *Rheticus's Treatise*, pp. 17–19.

50. Hooykaas's *Commentary* is found on pp. 102–46. This is not to underestimate the considerable value of Hooykaas's interpretation. It is replete with historical background and sensitive readings of Rheticus. My own interpretation would not have been possible without Hooykaas's groundbreaking work.

51. Hooykaas's *Commentary*, p. 105.

52. Translated by Edward Rosen, *Three Copernican Treatises* (New York: Dover, 1939), pp. 186, 187. See also Alexandre Koyré's discussion in *La Revolution Astronomique* (Paris: Hermann, 1961), pp. 31–35 and notes 91–95.

53. See Burmeister, *Rheticus*, vol. 1, pp. 40, 41.

54. *Praefatio in arithmeticen* (1536), Burmeister, *Rheticus*, vol. 1, p. 30.

55. Inquiry into natural things should be "non affirmando sed quaerendo." Rheticus's statement of the limits is, "ea tamen quaerendi dubitatio Catholicae fidei metas non debet." *Rheticus's Treatise* (1:43:65).

56. Ibid., in secs. 3–7, trans. pp. 66–68.

57. "It is apparent that the Holy Spirit did not want to speak in the manner of the philosophers." Ibid., sec. 5, trans. p. 67.

58. Ibid., secs. 5 and 6 give the theological purpose behind the doctrine of creation, trans. pp. 67–68.

59. Ibid., secs. 4 and 5, trans. p. 67; sec. 60, trans. p. 99. See especially sec. 8, p. 68. "And we have also to take note of the fact that the Holy Spirit has not wished to compose a course of physics, but rather a rule of life, and to teach how we may be made children of God" sec. 47, trans. p. 93.

60. Hooykaas, *Commentary*, p. 103.

61. *Rheticus's Treatise*, sec. 9 (trans. mine).

62. Hooykaas, *Commentary*, p. 105.

63. "For it is written that one shall not diverge from the words of the Lord, either to the right or to the left, and that the Word itself has the force of

the demonstration, since it has been given to us by God." *Rheticus's Treatise,* sec. 2, trans. pp. 65, 66.

64. Ibid., sec. 1. See also sec. 9, "Scripture . . . openly teaches that one should proceed in handling of such matters, not in the way of affirmation but of inquiry."

65. Ibid., sec. 12, trans. p. 70.

66. Ibid., sec. 12, trans. pp. 70, 71.

67. Hooykaas, *Commentary,* pp. 107, 110.

68. *Rheticus's Treatise,* sec. 50, trans. 94: "For it is clear that by divine ordination he teaches that these bodies are kept in their path" (trans. mine).

69. Rheticus says "no sensible man should think that we have taken on the task of harmonizing the earth's motion to Holy Scripture rashly, or from love of sophistry" (*ut quilibet vir bonus sentiat nos non temere, aut sophisticae amore provinciam hanc accommodandi terrae mobilitatem ad sacram scripturam suscepisse*) *Rheticus's Treatise* (16:48:73).

70. Ibid., sec. 7, p. 68.

71. From the preface printed in Wittenberg, 1543. Translation from Pierre Duhem, *To Save the Phenomena,* p. 72.

72. A comparison of selected calculations performed by Copernicus and Reinhold, as well as historical influence, can be found in Owen Gingerich, "Erasmus Reinhold and the Dissemination of Copernican Theory," *Studia Copernicana* 6 (1973): 43–62, 123–25, and reprinted in *The Eye of Heaven: Ptolemy, Copernicus, Kepler* (New York: The American Institute of Physics, 1993), pp. 221–51.

73. A. Birkenmajer, "Le commentaire inédit d'Erasme Reinhold sur le 'De Revolutionibus' de Nicolas Copernic," *La science au seizième siècle* (Paris, 1960), pp. 171–77.

74. The major work on Maestlin has been Richard Jarrell's "The Life and Scientific Work of the Tübingen Astronomer Michael Maestlin," Ph.D. diss., University of Toronto, 1971, and "Maestlin's Place in Astronomy" *Physis* 17, Fasc., 1–2 (1975): 5–20, and "Astronomy at the University of Tübingen: The Work of Michael Maestlin," in Friedrich Seck, ed. *Wissenschaftsgeschichte um Wilhelm Schickard* (Tübingen: Mohr, 1981).

75. Maestlin's *Disputationes tres astronomicae et geographicae* (1592) is now lost. His famous *Epitome Astronomiae: qua brevi explicatione omnia, tam ad sphaericam quam theoricam eius partem pertinentia, ex ipsius scientiae fontibus deducta* (Tübingen: Gruppenbacchius, 1597) contains geographical appendices to Book I and Book IV.

76. Robert A. Kolb, "Historical Background of the Formula of Concord," in Wilbert Rosin and Robert Preus, eds., *A Contemporary Look at the Formula of Concord* (St. Louis: Concordia, 1978), pp. 12–87.

77. For the majoristic controversy, see Kolb, "Historical Background," pp. 26–29. Erwin Lueker, ed., *Lutheran Cyclopedia* (St. Louis: Concordia, 1975), rev. ed., pp. 512, 513, and p. 30 for Nicholas von Amsdorf.

78. For the relation between natural philosophy and the church, see Melanchthon's discussion of the purpose and use of *physica* in *Initia*, pp. 190–91.

79. The most extensive treatment in English of the period of classic Lutheran orthodoxy is Robert D. Preus, *The Theology of Post-Reformation Lutheranism*, 2 vols. (St. Louis: Concordia, 1970).

3. Geoheliocentrism and the Bible: Brahe, Peucer, and Rothmann

1. No adequate biography of Peucer exists. For a short sketch of his life, see Robert Kolb, *Caspar Peucer's Library: Portrait of a Wittenberg Professor of the Mid-Sixteenth Century* (St. Louis: Center for Reformation Research, 1976).

2. For Peucer's work on astronomy, see Robert S. Westman, "The Melanchthon Circle, Rheticus and the Wittenberg Interpretation of the Copernican Theory," *Isis* 66 (1975): 178–81. His lectures appeared mainly in *Elementa doctrinae de circulis coelestibus, et primo motu, recognita et correcta* (Wittenberg, 1553), and his most famous work was *Commentarius de praecipuis divinationum generibus* (1553).

3. Little is known of Rothmann's life, but see Lettie S. Multhauf, "Christoph Rothmann," *DSB*, vol. 11, pp. 561–62, and Bruce T. Moran, "Christoph Rothmann, the Copernican Theory, and the Institutional and Technical Influences on the Criticism of Aristotelian Cosmology," *Sixteenth Century Journal* 13 no. 3 (1982): 85–108.

4. The literature on Rothmann's astronomical work is scarce, but see recently Bernard R. Goldstein and Peter Barker, "The Role of Rothmann in the Dissolution of the Celestial Spheres," *Historical and Philosophic Studies*, Princeton University, 1992.

5. *Brahe to Rothmann*, 21 February 1589, *Tychonis Brahe Dani Opera Omnia*, 15 vols., ed. J. L. E. Dreyer (Copenhagen: Nielsen & Lydiche, 1913), hereafter cited as *TBOO*. See vol. 6, p. 176. On Tycho's restoration project, see also Ann Blair, "Tycho Brahe's Critique of Copernicus and the Copernican System," *Journal of the History of Ideas* 51 (1990): 355–77, esp. p. 357.

6. On the "Tychonic Compromise," see J. L. Dreyer, *Tycho Brahe: A Picture of Scientific Life and Work in the Sixteenth Century* (1890; reprint by Dover, 1963); idem, *A History of Astronomy from Thales to Kepler* (1953; Dover,

2nd ed.); Dorothy Stimson, *The Gradual Acceptance of the Copernican Theory of the Universe* (New York: Baker & Taylor, 1917), p. 34; Victor E. Thoren, *The Lord of Uraniborg: A Biography of Tycho Brahe* (Cambridge: Cambridge University Press, 1991).

7. Owen Gingerich and Robert S. Westman, *The Wittich Connection: Conflict and Priority in Late Sixteenth-Century Cosmology* (Philadelphia: Transactions of the American Philosophical Society, 1988), vol. 78, part 7, p. 2.

8. See Tycho's comments on Reinhold in *Brahe to Peucer, TBOO,* 7, p. 137.

9. Chemical correspondences were common among the Paracelsians. See below, the literature in note 11 and the quotations from Tycho's writings in note 45.

10. *Brahe to Rothmann,* 17 August 1588, *TBOO,* 6, p. 144.

11. On Tycho's Paracelsianism and alchemical researches, see Alain Segonds, "Tycho Brahe et L'Alchimie" in Jean Claude Margolin and Sylvain Matton, *Alchimie et Philosophie à la Renaissance* (Paris: Librairie Philosophique, J. Vrin, 1993); Richard J. Shakelford, *Paracelsianism in Denmark and Norway in the Sixteenth and Seventeenth Centuries* (Ann Arbor: University Microfilms International, 1989); "Paracelsianism and Patronage in Early Modern Denmark" in Bruce T. Moran, *Patronage and Institutions: Science, Technology and Medicine at the European Court 1500–1750* (The Boydell Press, 1991); Owen Hannaway, "Tycho Brahe, Laboratory Design and the Aim of Science," *Isis* 84 (1993): 211–30. For a recent article that surveys the Scandinavian literature on Tycho's wide-ranging interests and program, see J. R. Christianson, "Tycho Brahe in Scandinavian Scholarship," *History of Science* 36 (1998): 467–484. J. R. Christianson's *On Tycho's Island* (Cambridge: Cambridge University Press, 2000) seeks to place Tycho's work in the social context of patronage in sixteenth-century Denmark.

12. As Gingerich and Westman point out, Tycho redefined the problem within the sphere of his competence as an observational astronomer. See Owen Gingerich and Robert S. Westman, *The Wittich Connection: Conflict and Priority in Late Sixteenth-Century Cosmology* (Philadelphia: Transactions of the American Philosophical Society, 1988), vol. 78, part 7.

13. "The question of celestial matter is not properly a decision of astronomers. The astronomer labors to investigate from accurate observations not what heaven is and from what its splendid bodies exist, but rather especially how all these bodies move. The question of celestial matter is left to the theologians and physicists among whom now there is still not a satisfactory explanation" *Brahe to Rothmann,* 17 August 1588 (*TBOO* 6:139:39–140:3).

14. See *TBOO,* 1, pp. 148, 149, 152 and 5, p. 5. The *Oratio* has been variously evaluated by the few historians of science who have paid any attention to it. Dreyer's straightforward description of its contents failed to offer any ex-

planation other than Tycho's imitation of "other astrological writers" J. L. E. Dreyer, *Tycho Brahe* (Edinburgh: Adam and Charles Black, 1891), pp. 74ff. Jardine regarded it as wholly unoriginal in almost every respect, similar to many other such addresses in the sixteenth century. See Nicholas Jardine, *The Birth of the History and Philosophy of Science* (Cambridge: Cambridge University Press, 1984). Since three-fourths of the document is concerned with astrology, Thoren saw its most prominent and controversial aspect as Tycho's defense of horoscope astrology. See Victor E. Thoren, *The Lord of Uraniborg: A Biography of Tycho Brahe* (Cambridge: Cambridge University Press, 1991), pp. 80–82.

15. The narrative concerning Seth is found in Genesis 4:25–5:7.

16. Abraham's sojourn in Egypt is narrated in Genesis 12:10–20.

17. Among the extensive literature on Paracelsianism, see Allen G. Debus, *The Chemical Philosophy* (New York: Science History Publications, 1977); on Severinus see especially vol. 1, pp. 130ff. Later in his lecture Tycho specifically attacked Thomas Erastus, who denied any connection between celestial bodies and medicine. Tycho's linkage of the seven planets to seven principal organs of human anatomy was no doubt inspired by Severinus's Paracelsian treatise. See *TBOO*, 1, pp. 166, 167.

18. Examples of biblical prohibitions can be found in Deuteronomy 13:1ff. and 18:14.

19. "The true uses of astrology are never censured in that [Mosaic] law but only its abuses," and "But one cannot infer from this that there is no astral influence or that it should not be done by humans" *TBOO*, 1, p. 162.

20. Tycho recorded conversations after the lecture in which Hemmingsen expressed reservations about Tycho's ideas. See ibid., pp. 170–73.

21. "So Moses discussing God's reason for making the heavens, the luminaries and the stars, says not only that they are for times, days and years but also for signs and they necessarily signify something to men for whose sake the greater part of this creation was made" ibid., p. 154.

22. 2 Chronicles 16:12.

23. *TBOO*, 1, pp. 162–63.

24. "So I will take it upon myself especially for those who take delight in these matters to adduce here some reasons that confirm the certitude of astrology" *TBOO*, 1, p. 153.

25. See Genesis 29:31–33:20; Malachi 1:6; Romans 9:10–13.

26. *TBOO*, 1, p. 162.

27. Malachi 1:6.

28. *TBOO*, 1, p. 162.

29. "So the heavenly bodies necessarily teach their meaning [*significatio*] by the power placed in them by God and so one can infer that there are causes

which signify. Nor does this in any way detract from divine omnipotence or liberty which are tied to secondary causes. Although God is a perfect and free agent, unrestricted by any natural law, yet he did not want to pervert the order of nature that he set up. And although God could have done everything without intermediaries . . . yet he was pleased by his inscrutable wisdom that all these things that normally happen in the world come from Him through means." *TBOO*, 1, p. 154.

30. *Brahe to Rothmann*, 21 February 1589, *TBOO*, 6, 178:3.

31. See Christine Schofield, *Tychonic and Semi-Tychonic World Systems* (New York: Arno Press, 1981), pp. 1–49.

32. Tycho hoped to write a work entitled *Theatrum Astronomicum* which he never completed. The first chapter of this work appeared in *Astronomiae Instauratae Progymnasmata* (1602).

33. "It was then that I perceived that the celestial appearances did not precisely agree with even the Copernican calculations, much less the Alphonsine [Ptolemaic] that had become obsolete." *Brahe to Peucer*, 13 September 1588, *TBOO*, 7, p. 128.

34. "Furthermore, Ptolemy assumes so many large epicycles which occupy a great amount of space and are superfluous. I judged that if everything can be explained with fewer devices, one should do so. I had great doubts whether there was any necessary cause or natural combination at hand to explain why the superior planets revolved around the sun so as to always occupy the summit of the epicycles at conjunction but were at their lowest point at opposition. I also wondered why the two inferior planets (i.e., Mercury, Venus) were always in the same mean position with regard to the sun and were conjoined in the apogee and perigee of the epicycles. It was not possible to arrive at a certain and evident system for this combination if indeed these stars were at so great intervals from one another to the point that this mutual concatenation would be in that precise manner to have some fitting necessity so that everything in it happened in this way. All this appeared less agreeable. Rather for this reason alone I suspected that something constrained, discordant and superfluous lay hidden in these assumptions." *Brahe to Peucer*, 13 September 1588, *TBOO*, 7, p. 128.

35. Ibid., p. 128.

36. Ibid., pp. 129, 130.

37. A complete history of the introduction of the solid spheres is yet to be written. Swerdlow claimed that the Latin translation of Ibn al-Haytham's *On the Configuration of the World* into Latin marked its beginnings in Western astronomy. See N. M. Swerdlow, "Pseudodoxia Copernicana," *Archives internationales d'histoire des sciences* 26 (1976): 108–58. See also Edward Rosen, "The

Dissolution of the Solid Celestial Spheres," *Journal of the History of Ideas* 46 (1985): 13–31. The most comprehensive discussion is to be found in William H. Donahue, *The Dissolution of the Celestial Spheres 1595–1650* (New York: Arno Press, 1981).

38. "Consequently, I did not doubt that my new system [*inventio*] was altogether rightly formed and that it offered nothing mathematically or physically absurd; I was led into the idea that the restoration [*redintegratio*] of the heavenly revolutions could be built up from both ancient and modern observations." *Brahe to Peucer*, 13 September 1588, *TBOO*, 7, p. 131.

39. *Brahe to Brucaeus*, 1584, *TBOO*, 7, p. 80.

40. Kepler examined the observations of 1582–83 but found no mention of parallax. He suggested from an examination of a manuscript that one of Tycho's assistants had calculated Mars's parallax from Copernicus and that Tycho had unknowingly adopted this as the original observations. Johannes Kepler, *De motibus stellae Martis*, *KGW* III: 121, ll. 18ff. Dreyer later confirmed that the manuscript in question was in Tycho's hand. See Owen Gingerich, "Dreyer and Tycho's World System," *Sky and Telescope* 64 (1982): 138–40. For discussions of Tycho's confusion, see J. L. E. Dreyer, *Tycho Brahe* (Edinburgh: Adam and Charles Black, 1891), pp. 178, 179 and Christine Schofield, *Tychonic and Semi-Tychonic Systems* (New York: Arno Press, 1981). Two works emphasize Tycho's need to prove his system. See Owen Gingerich and Robert S. Westman, *The Wittich Connection*, pp. 69–76, and Ann Blair, "Tycho Brahe's Critique of Copernicus and the Copernican System," *Journal of the History of Ideas* 51 (1990): 355–77.

41. Owen Gingerich and James R. Voelkel, "Tycho Brahe's Copernican Campaign," *Journal of the History of Astronomy* 29 (1998): 1–34.

42. Ibid., p. 11.

43. Gingerich and Voelkel discuss Tycho's flawed table of solar refraction on pp. 21–23.

44. See the *Oratio* in *TBOO*, 1, p. 172. Letters from Pratensis and Danzeus are partially translated and commented on in Schofield, *Tychonic and Semi-Tychonic World Systems*, pp. 34, 35.

45. "All structures of the human body are from heaven analogous to the properties of the seven wandering stars so that these almost similar operations share in our body with the nature of the planets in the heavens. Thus the heart in the human body, author of the vital spirit, compares to the sun from which proceeds a vivifying heat. Thus the brain is similar to the celestial moon, growing or decreasing like it, and is itself strengthened or lessened and rivals the moist nature of the moon and together with the moon is subject to constant unrest. But as the heart and brain are the two greatest members in our bodies, so also

the sun and moon that are analogous to these are the stronger luminaries in heaven. And as a clear intimacy exists between the heart and brain, so also there is a many-sided partnership between sun and moon." *Oratio De Disciplinis Mathematicis, TBOO,* 1, p. 157.

46. Owen Gingerich and Robert S. Westman, *The Wittich Connection,* p. 48.

47. *Brahe to Peucer,* 13 September 1588, *TBOO,* 7, p. 128. See also p. 129, "Nevertheless, the substantial absurdity of the earth's ordinary and continual revolution presented quite an obstacle to me, to say nothing of its being contrary to the unquestionable authority of the Sacred Scriptures."

48. *Rothmann to Brahe,* 13 October 1588, *TBOO,* 6, p. 149.

49. Ibid.

50. Ibid.

51. *Brahe to Rothmann,* 17 August 1588, *TBOO,* 6, p. 139.

52. *Brahe to Kepler,* 1 April 1598, *TBOO,* 8, p. 45.

53. "May we never admit something of an elementary nature (something that is corruptible, variable and conformed to this transitory region) to that pure, perfect and changeless celestial region. Since everyone has rightly and unanimously established that heaven is of a certain nature that is exempt from the number and nature of the four elements and far more excellent than them, they maintain that it has a certain kind of fifth essence far different from the four elements and far more excellent." *Brahe to Rothmann,* 17 August 1588, *TBOO,* 6, p. 135.

54. "But we disagree with you when you say that because of refraction there is a difference between the transparency (*diaphana*) of ether and air i.e. that there is no elementary air in the heavenly spheres but rather that ether is a liquid substance very different from elementary air," *Rothmann to Brahe,* 11 October 1587, *TBOO,* 6, p. 111, and "Nor would I ever have introduced the matter of air [into the celestial region] except that the failure of refraction and optical demonstrations compelled me." *Rothmann to Brahe,* 13 October 1588, *TBOO,* 6, p. 149.

55. *Oratio de disciplinis mathematicis, TBOO,* 1, p. 156.

56. See the quotation from the *Oratio de disciplinis mathematicis* in note 45 above.

57. Charles B. Schmitt, *Aristotle and the Renaissance* (Cambridge: Harvard University Press, 1983). See also Gingerich and Westman, pp. 72ff, and literature cited there for more leads.

58. Peter Barker, "Jean Peña and the Stoic Physics in the Sixteenth Century," *The Southern Journal of Philosophy* 23 (1985): 93–107.

59. *Brahe to Peucer,* 13 September 1588, *TBOO,* 7, p. 134.

60. Ibid., p. 135.

61. *Brahe to Rothmann,* 21 February 1589, *TBOO,* 6, p. 177.

62. *Brahe to Peucer*, 13 September 1588, *TBOO*, 7, p. 133.

63. *Brahe to Rothmann*, 21 February 1589, *TBOO*, 6, p. 177.

64. Ibid.

65. Ibid., 17 August 1588, *TBOO*, 6, p. 140.

66. For usage of *stereoma* in Greek philosophy, see Liddel and Scott, *Greek-English Lexicon*, ninth ed. (Oxford: The Clarendon Press, 1940), p. 1640.

67. *Brahe to Peucer*, undated letter of 1590, *TBOO*, 7, p. 231.

68. Castalio's Latin version of Isaiah 40:22 can be translated as "who extends the heavens like a membrane and expands it like a tabernacle for habitation," and Job 37:18, "did you lead ether with it so strongly that it appeared to be hard?"

69. *Brahe to Peucer*, 13 September 1588, *TBOO*, 7, p. 133.

70. *Peucer to Brahe*, 10 May 1589, *TBOO*, 7, p. 185.

71. *Brahe to Peucer*, 1590, *TBOO*, 7, p. 230.

72. *Peucer to Brahe*, 10 May 1589, *TBOO*, 7, p. 185.

73. "For my part I prefer to philosophize from Scripture only, than to talk nonsense with philosophers unskilled in divine things. This is especially true in matters where they stubbornly hang on to their opinions or are deeply blind. They sometimes invent or assume to be true, certain hypotheses which are contrary to the heavenly truth." *Peucer to Brahe*, 10 May 1589, *TBOO*, 7, p. 188.

74. *Brahe to Peucer*, 1590, *TBOO*, 7, p. 231.

75. Ibid., p. 230.

76. Ibid., p. 232.

77. Ibid., p. 235.

4. Kepler, Cosmology, and the Bible

1. Max Caspar's *Kepler*, trans. and ed. Doris Hellman (New York: Abelard-Schumann, 1959), is the standard biography. The greater part of Kepler scholarship has focused on the mathematical details of his astronomy, especially the announcement of the first two laws of planetary motion in the *Astronomia Nova*. This is no doubt what led Koyré (among others) to conclude that Kepler's central concern was: *à quo moventur planetae?* There is little doubt that planetary motion was important, but Kepler's polyhedral theory suggests that the size and distances of the planets were of equal concern. Recent Kepler scholarship has balanced the previous focus on the *Astronomia Nova* by paying more attention to his other works. A groundbreaking article was Gerald Holton's "Johannes Kepler's Universe: Its Physics and Metaphysics" reprinted

in *The Thematic Origins of Scientific Thought: Kepler to Einstein*, rev. ed. (Cambridge: Harvard University Press, 1988), pp. 53–74. For a recent exposition, see Bruce Stephenson, *Kepler's Physical Astronomy* (New York: Springer-Verlag, 1987). For a focus on Kepler's cosmology, see J. V. Field, *Kepler's Geometrical Cosmology* (Chicago: University of Chicago Press, 1988). Some wider concerns are also evident in Part 3 of Owen Gingerich, *The Eye of Heaven: Ptolemy, Copernicus, Kepler* (New York: American Institute of Physics, 1993). Only recently have adequate English translations of Kepler's works appeared. Among these are A. M. Duncan's translation of the *Mysterium Cosmographicum* as *The Secret of the Universe* (New York: Abaris Books, 1981), hereafter referred to as Duncan; William H. Donahue's translation of *Astronomia Nova* as *Johannes Kepler, New Astronomy* (New York: Cambridge University Press, 1992), hereafter referred to as Donahue; E. J. Aiton, A. M. Duncan, and J. V. Field, complete translation of *Harmonice Mundi* (Philadelphia: American Philosophical Society, 1995). All translations are mine unless otherwise noted. All citations of Kepler's works are from the standard edition Walter von Dyck and Max Caspar, *Johannes Kepler Gesammelte Werke* (Munich: C. H. Beck, 1938–), 20 vols. These will be cited as volume number in Roman numerals and page and line numbers in Arabic numerals (e.g., III: 35, ll. 22–24).

2. See the *Astronomia Nova* Introduction, III: 18–35 and Donahue, pp. 45–69.

3. See William H. Donahue, "Kepler's Cosmology," in Norriss Hetherington, ed., *Encyclopedia of Cosmology* (New York: Garland Press, 1993), pp. 346–53.

4. Edward Rosen, "Kepler and the Lutheran Attitude Towards Copernicanism in the Context of the Struggle Between Science and Religion," in *Vistas in Astronomy* 18 (1975): 317–37.

5. *Kepler to Maestlin*, 3 October 1595, XIII: 40, ll. 256–58. Translated in Owen Gingerich, "Kepler, Johannes," *DSB*, vol. 7, p. 308.

6. The recognition of how central Kepler's polyhedral theory is to his life's work is discussed in I. Bernard Cohen, Preface to Duncan, p. 9. Recognition of Kepler's religious beliefs is seen in Owen Gingerich, "Johannes Kepler," *DSB*, vol. 7, pp. 289–312. Kepler's theological views are treated in Jürgen Hübner, *Die Theologie Johannes Keplers zwischen Orthodoxie und Naturwissenschaft* (Tübingen: J. C. B. Mohr 1975).

7. Job Kozamthadam, S.J., *The Discovery of Kepler's Laws: The Interaction of Science, Philosophy and Religion* (Notre Dame: University of Notre Dame Press, 1994).

8. Owen Gingerich, "Johannes Kepler," *DSB*, vol. 7, p. 293.

9. *Prodromus dissertationum cosmographicarum continens Mysterium Cosmographicum de admirabili proportione deque causis coelorum numeri,*

magnitudinis motuumque periodicorum genuinis & propriis demonstratum per quinque regularia corpora Geometrica.

10. On the history behind the *Mysterium Cosmographicum* and the structure of its argument, see J. V. Field, *Kepler's Geometrical Cosmology* (Chicago: University of Chicago Press, 1988), pp. 30–72. Field also gives a comparative analysis of the 1596 and the 1621 editions, pp. 73–95.

11. "Yet so that the public body of mathematicians would not be entirely deceived, I wanted to warn them with my title that here was being offered not a mathematical work but a cosmographia. Mathematics are not treated with rigor just as in Aristotle's books *On the Heavens*, nor even really a genuine cosmographia but only a forerunner. From this men will understand that something more certain and better will be published." *Kepler to Maestlin*, March 1596, *KGW* XIII: 70, (nr. 32) ll. 32–37.

12. "leviores videtis ratiunculas, et levidensè quoddam cosmographicum" *Fragmentum orationis de motu terrae, KGW* XX: 148, ll. 10–18.

13. One may recall that Maestlin, following established tradition, had divided astronomy into the two parts *sphaerica* and *theorica* in his summary of the science. Michael Maestlin, *Epitome Astronomiae: qua brevi explicatione omnia, tam ad sphaericam quam theoricam eius partem pertinentia, ex ipsius scientiae fontibus deducta* (Tübingen: Gruppenbacchius, 1597).

14. For a discussion of the meaning of *mundus* in early modern Latin and Kepler, see Duncan, p. 14 and André Le Boeuffle, *Astronomie, Astrologie: Lexique Latin* (Paris: Picard, 1987), nr. 808, p. 187.

15. Ptolemy, *Geographia*, Book I, chapter II, translation by Simon Knight from the Italian edition of Lelio Pagani 1990.

16. See Kepler's annotations, *KGW* VIII: 15. Reprinted and translated in Duncan, p. 51.

17. Note how *physica* and *cosmographia* can be used interchangeably by Kepler. See *Mysterium Cosmographicam, KGW* I: 16, ll. 19–21.

18. In Greek, *kosmos* = world and *graphia* = writing, description.

19. *Mysterium Cosmographicum, KGW* I: 5.

20. "Quam vocem, coelis et Naturae rerum dum aperimus his pagellis clarioremque efficimus." *Mysterium Cosmographicum, KGW* I: 6.

21. "Mysterium autem pro Arcano usurpavi et pro tali venditavi inventum hoc: quippe in nullius Philosophi libro talia unquam legeram." See Latin text in Duncan, p. 50. My translation differs somewhat from Duncan's.

22. Kepler's wording may well mean that he had not read any such theory in the intervening twenty-five years between the first and the second editions, not necessarily that this was his intention in 1596. Indeed, this seems the natural interpretation since he refers to the fate of the book in those years in his remarks on *cosmographia* in the preceding sentence of his text.

23. Ephesians 5:22–33. The Pauline usage in turn draws on the nuptial imagery of the Hebrew Bible to characterize the relation of Yahweh to Israel. See Hosea 1 and 2. The Song of Songs was interpreted by both Jewish rabbis and Christian teachers as an allegory of God and Israel (or the church).

24. Many church Fathers, both eastern and western, wrote on the sacraments. Though *mysterion* originated in the Greek-speaking eastern Roman Empire, its borrowed form *mysterium* quickly found its way into the Latin churches of the West. See Ambrose of Milan, *De mysteriis*, circa a.d. 390 (*PL* edition, vol. 16, pp. 405–26).

25. For Kepler's doctrine of the Eucharist, see Jürgen Hübner, *Die Theologie Johannes Keplers zwischen Orthodoxie und Naturwissenschaft* (Tübingen: J. C. B. Mohr, 1975), pp. 3ff., 23, 37ff., 115ff., 121ff., 138ff.

26. *Harmonice Mundi*, Book V, Preface, *KGW* VI: 289.

27. Ibid., 290. Among the several passages referring to the Israelites plundering the Egyptians, see Exodus 3:21–22.

28. A well-known instance is found in Augustine's *De Doctrina Christiana*, Book II.

29. Introduction, *Astronomia Nova*, *KGW* III: 19.

30. Kepler does refer to this distinction in his discussion of Ecclesiastes 1:4. "You do not hear any physical dogma here. The message is a moral one, concerning something self-evident and seen by all eyes but seldom pondered." Introduction, *Astronomia Nova*, *KGW* III: 31, ll. 22–23. Donahue's translation, p. 63. However, this statement does not capture the full range of Kepler's use of the Bible.

31. Jürgen Hübner, *Die Theologie Johannes Keplers zwischen Orthodoxie und Naturwissenschaft* (Tübingen: J. C. B. Mohr, 1975), pp. 161, 163.

32. Jacob Heerbrand, *Compendium Theologiae* (1579), p. 25, "Let the comparison of Scripture be with similar things and explain the more obscure things by the clearer." Matthias Haffenreffer, *Loci Theologici* (4th ed.; 1609), quotes Augustine *De Doctrina Christiana*, Book II, chap. 9 and Book III, chap. 26. See Hübner, p. 163, n. 17 and p. 164, n. 18. I have translated the Latin quotations in Hübner's notes.

33. For Lutheran statements on the sufficiency of Scripture, see Haffenreffer, *Loci Theologici*, p. 125 and Hübner, p. 164, n. 19.

34. Cf. Kepler, "maneamus primum in apertis dictis Scripturae," *KGW* XVII (nr. 808, letter): 51. See Hübner, p. 161, n. 2. Could it be a statement of limitation? "Let's limit ourselves to the clear statements of Scripture."

35. Matthias Haffenreffer, *Loci Theologici*, "Scripture is sufficient and perfect for defining and determining those things that are necessary for the knowledge of God and his will, for the observance of divine worship, and the obtaining of eternal life." (See Hübner, p. 164, n. 19.) Also, "I judge that Sacred

Scripture which God revealed for our edification and salvation is plain, simple and perspicuous and that whatever is necessary for the knowledge of our salvation and the exercise of piety can be well known if one is a pious and attentive reader." (See Hübner, p. 164, n. 20.) Translations are mine.

36. The once standard picture of the separate worlds of Renaissance humanism and natural philosophy in which the latter eclipsed the former—so emphasized by Bacon and Descartes, and promoted by twentieth-century historians—has found considerable modification in recent years. See Anthony Grafton, "Humanism and Science in Rudolphine Prague: Kepler in Context," in *Defenders of the Text: Traditions of Scholarship in an Age of Science, 1450–1800* (Cambridge, Mass.: Harvard University Press, 1991).

37. For example, Henry Savile, *Prooemium mathematicum* (1570).

38. Kepler expressed and developed his views on the history of astronomy in *Apologia pro Tychone contra Ursum* (1601) and later in his last great work, *Tabulae Rudolphinae* (1627). For discussion of Kepler's work on the history of astronomy, see Nicholas Jardine, *The Birth of History and Philosophy of Science: Kepler's A Defence of Tycho against Ursus, with Essays on Its Provenance and Significance* (Cambridge University Press, 1984), pp. 258–86 and Grafton, "Humanism and Science in Rudolphine Prague."

39. Anthony Grafton, "Kepler as a Reader," *Journal of the History of Ideas* 53 (4): 561–72.

40. This was the most widely read section of Kepler's writings for over three centuries and probably influenced many who attempted to reconcile Copernicanism and the Bible. See Owen Gingerich, "Foreword," in Donahue, p. xii.

41. Introduction, *Astronomia Nova*, *KGW* III: 33. Also, Donahue, p. 66.

42. Psalm 19, Psalm 104; Joshua 10:12; Ecclesiastes 1:4.

43. Genesis 1; Jeremiah 31:37; Job 38; Psalms 24 and 137.

44. Kepler says that the psalmist wrote a hymn to God, not a speculation about physical causes. "Atqui longissime abest Psaltes a speculatione causarum Physicarum. Totus enim acquiescit in magnitudine Dei qui fecit haec omnia, Hymnumque pangit Deo conditori . . ." Introduction to the *Astronomia Nova*, *KGW* III: 31. It is worth noting that Kepler chose the verb *pango*, a verb in traditional Christian usage, to denote adoration of God, e.g., Thomas Aquinas's *Pange Lingua*.

45. Hübner (p. 169) wrongly attributes to Kepler the modern view that Genesis 1 is a poetic expression of doxology with little historical value, citing Claus Westmann, *Genesis Biblische Kommentar*, vol. 1 (1968). From Kepler's belief that Psalm 104 was poetry and that Genesis 1 had a poetical structure, it does not follow that he believed it was unhistorical.

46. He treats in order: Psalm 19; Job 10:12; Psalm 24; Psalm 137; Ecclesiastes 1:4; Psalm 104.

47. "If he were [an astronomer], he would not fail to mention the five planets, than whose motion nothing is more admirable, nothing more beautiful, and nothing a better witness of the Creator's wisdom, for those who take note of it" *Astronomia Nova, KGW* III: 32, ll. 32–41. Donahue's trans., p. 65.

48. *Kepler to Herwart,* 28 May 1605, *KGW* XV: 182.

49. Kepler at times invokes the Bible in a positive fashion to refute false positions in natural philosophy. He was convinced that the universe could be neither eternal nor infinite. The *Epitome of the Copernican Astronomy* engages Aristotle's opinions about the nature of the heavens. The Preface to the Reader of Book IV answers a number of objections which are designed to capture the support of his patron, Rudolph II. One of his most prominent strategies of argumentation counters Aristotle's notion of the eternity of the world. Kepler is intent on demonstrating that the universe was created in time, a goal which constitutes one of the main aims of his building "the edifice of astronomy." His theories will provide at once the refutation of Aristotle and the display of the Creator's glory (*Epitome,* Book IV, pp. 846, 847, 848). Yet, while astronomy provides such a refutation, Christ himself confirms the mutability and therefore temporality of the universe. The heavens will pass away and be destroyed. How could Aristotle be right when the authority of Christ speaks thus? ("since not Tycho, not I, but Christ Himself pronounces concerning the visible world" *Epitome,* Book IV, p. 848.)

50. *Kepler to Herwart,* 25 March 1598, *KGW* XIII, nr. 91, ll. 182–84.

51. See Kepler's notes to Matthias Haffenreffer's *Letter to Kepler,* 31 July 1619, *KGW* XII: 29, ll. 1–10.

52. See *Astronomia Nova* chapter summaries, *KGW* III: 36, ll. 4–15; Donahue, p. 78. *Epitome Astronomiae Copernicanae* (1618), Book IV, *KGW* VII: 254, ll. 11–29.

53. "Ut Copernicus Mathematicis, sic ego Physicis, seu mauis, Metaphysicis rationibus" *Mysterium Cosmographicum,* Preface to the Reader, *KGW* I: 9.

54. *Kepler to Herwart,* 10 April 1599, *KGW* XIII, nr. 117, ll. 295–96.

55. Kozhamthadam lists eight occurrences of the sphere icon of the Trinity. See Job Kozhamthadam, *The Discovery of Kepler's Laws: The Interaction of Science, Philosophy and Religion* (Notre Dame: University of Notre Dame Press, 1994), p. 17.

56. *Kepler to Maestlin,* 3 October 1595, *KGW* XIII: 35, ll. 71–74.

57. *Epitome Copernicanae Astronomiae,* Book I, part 2, *KGW* VII: 51, ll. 1–22. Kepler here draws on the language of the Western form of the Nicene Creed which asserts that the Spirit proceeds from the Father *and the Son (filioque).*

58. ". . . ad illorum naturam coelorum numerum, proportiones, et motuum rationem accommodaverit," Preface to the Reader in *Mysterium Cosmographicum, KGW* I: 9.

59. "Quod si (cogitabam) Deus motus ad distantiarum praescriptum aptavit orbibus, utique et ipsas distantias ad alicuius rei praescriptum accommodavit," Preface to the Reader in *Mysterium Cosmographicum, KGW* I: 10.

60. Preface to the Reader in *Mysterium Cosmographicum, KGW* I: 11.

61. *Mysterium Cosmographicum,* chap. 22, *KGW* I: 75–77.

62. William H. Donahue, "Kepler's Invention of the Second Planetary Law," *The British Journal for the History of Science* 27 (1994): 95, translated by Donahue. The original Latin is found in note 15.

63. *Astronomia Nova, KGW* III: 20, ll. 18–22; Donahue, p. 48.

64. ". . . bear with me now patient reader, if I trifle for a moment with a serious subject, and indulge in allegories a little. . . . if this meets with less approval from the reader, let him reflect, that this is a by-product, and not the main point." *Mysterium Cosmographicum, KGW* I: 30, 31. The translations in this paragraph are Duncan's unless otherwise noted (p. 107). Kepler's words here should not be taken as indicating a form of play that lessens the significance of his point. Kepler elsewhere says that God plays with us by showing the Trinity in the sphere, but he does not at all mean that this should not be taken seriously. In the second edition of the *Mysterium,* chapter 4 is one of the chapters in which he made no changes or modifications.

65. *Mysterium Cosmographicum, KGW* I: 30, 31; Duncan, p. 107.

66. Ibid.

67. *Harmonice Mundi,* Book V, chap. 9, *KGW* VI: 330, ll. 30ff.

68. These features are usually headed by subjunctive verb forms expressing Kepler's wishes. The biblical allusions include: "imitators of God" (Eph. 5:1); "[God] chose the Church and cleansed it from sin by the blood of His Son" (Eph. 5:26); Kepler's catalogue of sins is reminiscent of Paul's in Ephesians 4:31 and the phrase "works of the flesh" comes from Galatians 5:19; "in deeds . . . making sure of their calling" (2 Pet. 1:10; it seems Kepler read the Vulgate text).

5. Copernican Cosmology, Cartesianism, and the Bible in the Netherlands

1. There is a paucity of histories of Dutch science. Among recent works, see H. A. M. Snelders and K. van Berkel, eds., *Natuurwetenschappen van Renaissance tot Darwin* (Den Haag, 1981); Dirk Struik, *The Land of Stevin and Huygens: A Sketch of Science and Technology in the Dutch Republic during the Golden Century* (Dordrecht, 1981); Klaas van Berkel, *In het voetspoor van*

Stevin. Geschiedenis van de natuurwetenschap in Nederland 1580–1940 (Meppel, 1985). For a review, see Floris Cohen, "Open and Wide, Yet Without Height and Depth," *Tractrix* 2 (1990): 159–65. On the influence of Italian science in the Netherlands, see C. S. Maffioli and L. C. Palm, eds., *Italian Scientists in the Low Countries in the XVIIth and XVIIIth Centuries* (Amsterdam, 1989). Also see Klass van Berkel, Albert van Helden, and Lodewijk Palm, eds., *A History of Science in the Netherlands* (Leiden: Brill, 1999).

2. Biographical information on Frisius can be found in A. De Smet, "Gemma Frisius" in *Nationaal Biographisch Woordenboek,* vol. 6 (Brussels, 1974), columns 315–31.

3. See Leuven, Stedelijk Museum, *550 Jaar Universiteit Leuven* (Lembeke, 1976), pp. 212–213. On Frisius's early exposure to the Copernican system, see E. H. Waterbolk, "The Reception of Copernicus's Teachings by Gemma Frisius (1508–1555)" *Lias* 1 (1974): 225–42.

4. The relevant portion of the text of the *Epistola* is reproduced in Grant McColley, "An Early Friend of the Copernican Theory: Gemma Frisius" *Isis* 26 (1936): 322–25.

5. "I omit [mentioning] the intolerable errors in the motion of Mercury" ibid.

6. "Although Ptolemy's hypotheses are on the surface more plausible than Copernicus's, nevertheless they involve some absurd things, both when stars are taken to move with unequal motions in their circles and when they don't have any evident causes for the motions. For Ptolemy assumes that when the three superior planets are acronuchoi (or position from the diameter of the sun), they are in the perigee of their epicycle and he accepts this as a fact (to hoti). Copernicus's hypotheses teach the same as necessary and they demonstrate it by giving reasons (to dioti)." Ibid., p. 324.

7. See the description of Frisius's 1530 work entitled *De principiis astronomiae et cosmographiae deque usu globi* in *550 Jaar Universiteit Leuven*, p. 213.

8. Surprisingly little has been written on Beeckman. The most comprehensive treatment can be found in Klaas van Berkel, *Isaac Beeckman (1588–1637) en de mechanisering van het wereldbeeld* (Amsterdam, 1983). See especially chapter 6 for a discussion of religion and science in Beeckman's thought.

9. Stevin's biography is treated in E. J. Dijksterhuis, *Simon Stevin* (The Hague, 1943) and Rudolf Grabow, *Simon Stevin* (Leipzig, 1985).

10. For an exposition of Stevin's world system, see the introduction to *van de Hemelloop* in A. Pannekoek, ed., *The Principal Works of Simon Stevin* (Amsterdam: C.V. Swets & Zeitlinger, 1961), vol. 3, pp. 5–21.

11. The reception of Copernicanism in the Netherlands has not been adequately handled but is briefly treated in R. Hooykaas, "The Reception of Copernicanism in England and the Netherlands" in *The Anglo-Dutch Contri-*

bution to the Civilization of Early Modern Society (London, 1976), pp. 33–44 and Rienk Vermij, "Het copernicanisme in de Republiek: een verkenning" in *Tijdschrift der geschiednis* (1993).

12. Calvin's interpretative method is not our interest here, but helpful research may be found in the following: Richard C. Gamble, "*Brevitas et Facilitas*: Toward an Understanding of Calvin's Hermeneutic," *Westminster Theological Journal* 47 (1985): 1–17; Gamble, "Exposition and Method in Calvin," *Westminster Theological Journal* 49 (1987): 153–165; Hans Joachim Kraus, "Calvin's Exegetical Principles," *Interpretation* 31 (1977): 8–18; Thomas F. Torrance, *The Hermeneutics of John Calvin*, Monograph Supplements to the Scottish Journal of Theology (Edinburgh, 1988); Pieter A. Verhoef, "Luther's and Calvin's Exegetical Library," *Calvin Theological Journal* 13 (1968): 5–20; David Wright, "The Ethical Use of the Old Testament," *Scottish Journal of Theology* 36 (1983): 463–85.

13. "*Philosophos caecutisse dicemus cum in exquisita ista naturae contemplatione, tum artificiosa descriptione.*" *Institutio Christianae Religionis*, book 2, ch. 2, sec. 15. Peter Barth and Wilhelm Niesel, eds., *Johanis Calvini Opera Selecta* vol. III (Munchen, 1926), p. 258 and "*equidem qui liberales illas artes vel imbiberunt, vel etiam degustarunt, earum subsidio adiuti, longe altius provehuntur ad introspicidenda divinae sapientiae arcana*" book 1, ch. 5, sec. 2, *Johanis Calvini Opera Selecta* vol. III (Munchen, 1926), p. 46.

14. Calvin, *Corpus Reformatorum* XXXVI, p. 483, quoted in Torrance, op. cit., p. 194, n. 362. A translation of *De Scandalis* can be found in *Concerning Scandals* translated by John W. Fraser (Grand Rapids, 1978).

15. The *christiana philosophia* was spoken of in the *Concio Academica*. "If delight of the mind and rest from cares is sought that has in view living well and happy, then christian philosophy abundantly supplies" in Barth and Niesel, eds., *Johanis Calvini Opera Selecta* vol. 1, pp. 4–10. Most modern Calvin scholars do not think Calvin wrote the *Concio Academica*, which Nicolas Cop addressed to the faculty of the University of Paris on 1 November 1533. Even so, it seems likely that the views expressed in the address were also Calvin's since a copy of it was found in his handwriting in 1565 by Colladon. Calvin's deep interest in promoting the liberal arts is seen also in his instrumental role in the founding of the Genevan Academy which was begun on 16 March 1559, something that represented one of Calvin's greatest legacies to the city of Geneva. See W. Stanford Reid, "Calvin and the Founding of the Geneva Academy," *Westminster Theological Journal* 18 (1955): 1–33.

16. For Calvin's use of Aristotelian natural philosophy, see Christopher Kaiser, "Calvin's Understanding of Aristotelian Natural Philosophy: Its Extent

and Possible Origins" in Robert V. Schnucker, ed., *Calviniana: Ideas and Influences of Jean Calvin*, vol. X, Sixteenth Century Essays & Studies (1988).

17. "Astronomers . . . investigate with great labor whatever the keenness of man's intellect is able to discover . . . The study of astronomy not only gives pleasure but is also extremely useful. And no one can deny that it admirably reveals the wisdom of God . . . Clever men . . . ought not to neglect work of that kind." Commentary on Genesis 1:16 in *Corpus Reformatorum* XXIII, 22; translation in Calvin's *Commentaries*, ed. Joseph Haroutunian and Louise Smith (London: Library of Christian Classics), p. 356.

18. "If the astronomer inquires respecting the actual dimensions of the stars, he will find the moon to be less than Saturn; but this is abstruse, for to the sight it appears differently. Moses therefore adapts his discourse to common usage." Calvin's *Commentaries*, p. 356.

19. Calvin, *Commentary on Psalms*, vol. 5 (Grand Rapids, 1963), pp. 184–85 (first English version 1571).

20. Rosen outlined how the popular but spurious quotation from Calvin found currency. See Edward Rosen, "Calvin's Attitude toward Copernicus," *Journal for the History of Ideas* 21 (1960): 431–41. The debate began with Rosen's article. Various positions were taken in a stream of articles which followed. See Joseph Ratner, "Some Comments on Rosen's 'Calvin's Attitude toward Copernicus,' " *Journal for the History of Ideas* 22 (1961): 382–88; Pierre Marcel, "Calvin et la science: Comment on fait l'histoire," *Revue reformée* 69 (1966): 51; B. A. Gerrish, "The Reformation and the Rise of Modern Science" in Jerald C. Brauer, ed., *The Impact of the Church upon Its Culture* (Chicago, 1968), pp. 231–65; Richard Stauffer, "Calvin et Copernic," *Revue de l'histoire des religions* 179 (1971): 31–40; Robert White, "Calvin and Copernicus: The Problem Reconsidered," *Calvin Theological Journal* 15 (1980): 233–43; Christopher Kaiser, "Calvin, Copernicus and Castellio," *Calvin Theological Journal* 21 (1986): 5–31.

21. Kaiser, "Calvin, Copernicus and Castellio," p. 31.

22. Ibid., p. 6.

23. Ibid.

24. "non sic quotidie caeco naturae instinctu solem oriri et occidere quin ipse ad renovandam paterni erga nos sui favoris memoriam cursum eius gubernet" *Institutio Christianae Religionis* book 1, ch. 16, sec. 2. Barth and Niesel, eds., *Johanis Calvini Opera Selecta* vol. III, p. 190.

25. Quoted in Hans-Joachim Kraus, "Calvin's Exegetical Principles," *Interpretation* 31 (1977): 8–18. The original is from *Corpus Reformatorum* vol. 38, p. 405.

26. The text of Calvin's speech can be found in B. J. Kidd, ed., *Documents of the Contintental Reformation* (London, 1967), pp. 549–54. Of the nine quotations from the fathers, six were from Augustine. See the discussion of the

Lausanne disputation by William N. Todd, *The Function of the Patristic Writings in the Thought of John Calvin*, dissertation at Union Theological Seminary, New York, 1964, pp. 96ff.

27. *Reply to the Letter of Cardinal Sadleto to the Senate and People of Geneva* (1539) in John Dillenberger, ed., *John Calvin: Selections from His Writings* (Missoula, 1975), p. 92. "Will you obtrude upon me, for the Church, a body which furiously persecutes everything sanctioned by our religion, both as delivered by the oracles, and embodied in the writings of the Holy Fathers and approved by ancient Councils," p. 93.

28. "Nor ought you [Sadleto] to charge our doctrine [of the Eucharist] with novelty, since it was always held by the Church as an acknowledged point . . . the better course will be for you to read Augustine's *Epistle to Dardanus*, where you will find how one and the same Christ more than fills heaven and earth with the vastness of his divinity and yet is not everywhere diffused in respect of his humanity." Dillenberger, *John Calvin: Selections*, p. 99.

29. *Institutio Christianae Religionis* book 4, ch. 17, sec. 21. Barth and Niesel, *Johanis Calvini Opera Selecta*, vol. V, p. 371.

30. Robert H. Ayers, "The View of Medieval Biblical Exegesis in Calvin's *Institutes*," *Perspectives in Religious Studies* 7 (1980): 188–93.

31. On the role of the Emden synod in Dutch Calvinism, see Andrew Pettegree, *Emden and the Dutch Revolt: Exile and Development of Reformed Protestantism* (Oxford, 1992).

32. For the role of religion in urban life, see J. L. Price, *Holland and the Dutch Republic in the Seventeenth Century: The Politics of Particularism* (Oxford, 1994), pp. 70–80.

33. The full titles are: *Bedenckingen Op den Dagelijckschen, ende Iaerlijckschen loop van den Aerdt-kloot. Mitsgaders Op de ware Af-beeldinge des sienelijcken Hemels; daer in den wonderbare Wercken Godts worden ontdeckt* (Middelburg, 1629) and the Latin translation by Hortensius, *Commentationes in motum terrae diurnum et annum et in verum adspectabilis Coeli typum, in quibus epistymonikos osenditur diurnum annumque motum, qui apparet in sole et coelo sed soli terrae; simulque adspectabilis Coeli typus ad vivum exprimitur* (Middelburg, 1630). The same treatise appeared in a French translation in 1633. The original Dutch version was unavailable to me; all citations and translations from this treatise are from the Latin version as printed in Lansbergen's *Opera Omnia* (1663).

34. *Progymnasmatum astronomiae restitutae* (1619). His work on chronology appeared under the title *Chronologiae Sacrae libri sex, in quibus annorum mundi series, ab orde condito, ad eversa per Romanos Hierosolyma, nova methodo ostenditur* (Amsterdam, 1624; Middelburg, 1645). A year after Lans-

bergen's theological defense, he issued two more astronomical works: *Uranometriae libri tres, in quibus lunae solis et reliquorum planetarum et inerrantium stellarum distantiae a terra magnitudnes hactenus ignotae perspicue demonstratur* (Middelburg, 1631) and *Triangulorum Geometricorum libri quattuor* (Middelburg, 1631).

35. *Commentationes* in Lansbergen's *Opera Omnia* (1663), fol. 34.

36. Ibid., fol. 3.

37. "He [Copernicus] wanted to declare the causes of motion because we finally know something when we know its causes." Ibid., fol. 1.

38. "We will solidly show that nothing should be received in this science [*ars*] except what can be established with the firmest reasons . . . I will consider the Copernican hypotheses that the truth may more clearly shine from a mutual comparison [*oppositione*]." Ibid.

39. "[Copernicus] follows the order of nature when he attributes the most rapid motion of twenty-four hours to the earthly equinox because there is no smaller circle in the heavens. . . . Ptolemy attributes an equally fast motion to the eight spheres which are of diverse magnitudes. This is absurd because God made the motion of each sphere proportional to its magnitude." Lansbergen's *Commentationes*, fol. 2.

40. "God the Almighty testifies that 'he made all things in weight, number and measure' (Wisdom 11:22). From this it becomes clear that the position of Ptolemy is false, indeed impossible. Copernicus on the contrary established it as true that the earth moves from west to east each day because it is possible. The terrestrial equinox contains 5400 German miles which divided by 24 yields 225 miles for one hour of earth's motion. From this in the space of one second, three miles can be transversed . . . So the Copernican hypothesis is in every way consistent with nature." Ibid., fol. 2; "But it is evident that the earth is not at rest, rather it experiences a constant movement because it consists of the four elements which operate together on one another by the plan of God [*ex Deo instituto*]. This is a place of change and its appearance changes at each moment. The great Gentile teacher confirms the testimony of 1 Cor. 7:31 that 'the shape of this world is passing away.'" Ibid.

41. Lansbergen often cited the Pauline expression, "God is a God of order, not of confusion," (1 Cor. 14:33) as the basis of his argument for order in nature. See *Commentationes*, fol. 20.

42. *Commentationes*, fol. 6.

43. R. Hooykaas, *Rheticus's Treatise on Holy Scripture and the Motion of the Earth* (Amsterdam, 1984), pp. 178–81.

44. *Commentationes*, fol. 6.

45. Ibid., fol. 7.

46. Ibid.

47. Ibid.

48. Lansbergen's treatment of the true world system begins on fol. 16, *Commentationes.*

49. Lansbergen, *Commentationes,* fol. 18.

50. Ibid.

51. "It is apparent that the celestial spheres have been put together by this order of God Almighty so that if you invert or change anything, you will destroy and violate the total form of the heavens." Ibid., fol. 20.

52. Ibid., fol. 21.

53. "At least in the human body there are three faculties which all have the same function i.e. the vital which is the heart, the natural which is in the liver and the intellectual which is in the brain." Ibid., fol. 7.

54. "For when God created the earth for the sake of man, he gave it for his habitation (Ps. 115:16); also in ordering the sphere he had in view the globe of the earth and man. But if he had placed it [the earth] in the center of the first heaven, the place of the sun, it would have removed man the farthest from heaven, his throne. But [neither could he do it] in the sphere of Saturn or then it would have been removed too much from the sun or neighboring planets. So he placed it in the middle of the planets, the most dignified place of the first heaven." Ibid., fol. 22.

55. "This is the highest beneficence of God toward man where, although he made him a little lower than the angels, he crowned him with glory and honor (in the most prominent place in the first heaven)." Ibid.

56. "In the diurnal God made man a participant not only in the light and heat of the sun but also in its *life-giving power* . . . Then also the earth moves through the ecliptic each year. We should conclude that God Almighty carries us around through the ecliptic in a triumphal chariot so that every year we may gratefully praise and celebrate the Author with pleasure by a admirable inspection of his works." Lansbergen, *Commentationes,* fol. 22.

57. On the history of the University of Louvain, see Leuven, Stedelijk Museum, *550 Jaar Universiteit Leuven* (Lembeke: J. Roegiers, 1976). In this chapter I shall use the French form of both Froimond's name and the university where he taught.

58. See R. J. Hooykaas, *Rheticus's Treatise,* pp. 181ff.

59. Froimond's later theological work revolved around Jansenism and influenced the Catholic Diocese of Utrecht to a great degree. While Froimond was still a professor in Louvain, the Calvinist theologian of Utrecht, Gisbert Voet, wrote his *Causae Desperatae* in which he attempted to refute Jansen's claims to be a true follower of Augustine. Froimond responded with *Crisis*

Causae Desperatae (1636) and was soon afterwards given a position on Voet's home ground in Utrecht for the express purpose of further refuting the Calvinists. In Utrecht, Froimond issued an open letter to Voet (entitled *Syco-phantia Epistola ad Gisb. Voetium* 1640) attempting to refute the Reformed doctrines which the Catholic Church considered departures from an orthodox interpretation of Augustine. In the end, the Vatican also condemned Jansen's interpretation of Augustinianism, no doubt due to its affinities with the Calvinist heresy.

60. "The rest of the earth can be understood by this argument. Artillery is shot in tranquil air by the same powder, by the same globe, at the same elevation under the same parallel to east as to west. If the earth were rotating to the east, I say that any future shot would be far shorter to the east than to the west. If we measure its distance on the earth's surface, experience shows them both equal." Libert Froimond, *Anti-Aristarchus* (Louvain, 1631), p. 49.

61. "A ball itself shot from a bronze cannon equals the speed of the Copernican earth. Yet we have shown from Tycho and Kepler (as Lansbergen himself admits) that a point of the earth under the equator travels in one minute of an hour a little less than 4 German miles. Which time, in the experiment of the Landgrave and Rothmann that Tycho and Kepler endorsed, the ball of a cannon wanders (pererro) only about 1/2 German mile (i.e. 2500 pass.). Or certainly, as others, tell with a greater force 3,000 pass. i.e. moderately one Belgian mile. But I think that Lansbergen exaggerates the speed of a ball and increases it eight times." Ibid.

62. Libertus Fromondus, *In Epistolam Catholicam Beati Jacobi Apostoli*, reprinted in *Scripturae Sacrae Cursus Completus* ed. J. P. Migne, vol. 25 (Paris, 1840), pp. 683–90.

63. Froimond, *Anti-Aristarchus*, p. 39.

64. Ibid., p. 30.

65. Ibid.

66. "In the heat of battle did Joshua have a lapse of language when he commanded the sun to stand? . . . Yet what shall we do about the Holy Spirit who said, 'sun and moon stood' and again 'the sun stood in the middle of the sky' like an object of wonder 'it did not hasten to set?'" Froimond, *Anti-Aristarchus*, pp. 31, 32.

67. "So when Scripture says that the diameter of a vessel 'from lip to lip' was 10 cubits, it understands this to be a round number of a cord or [when] it says of a certain circle which encompasses the periphery of the lip that it is 30 cubits, it is true, even if it ought to be a little longer, i.e. 31 cubits. But if the periphery is precisely 30 cubits, the diameter is different from a length of 10 cubits by a particular magnitude [magna]. So it is true that the precise Archimedean

proportion of the diameter to the circumference cannot be drawn from a text of Scripture." Ibid., pp. 37, 38.

68. "But something mechanical and undeveloped can be [drawn from Scripture], so that this dart can be thrown back at Lansbergen who shuns any geometrical use from Sacred Scripture. Something can be drawn from this text, in which he struggles to show the inutility of Scripture, if we don't look for precision but something near precision." Ibid., p. 38.

69. Ibid.

70. "Again therefore, it would be unusual for Scripture to use words that normally signify local motion and to explain these phrases contrary to the common voice and interpretation of the Holy Fathers as not pertaining to truth but to appearances. This is what prudent reason and Trent forbids." Ibid., p. 31.

71. "From St. Augustine's error in a natural matter in using Scripture can we conclude that nothing can be drawn from Scripture concerning nature? Look how the heretics have deduced their perverse dogmas from sacred Scripture by their twisted logic: is it [Scripture] therefore useless for bringing out Catholic dogma?" Ibid., p. 37.

72. "St. Augustine's alleged weak-mindedness is in fact twisted. For his error is far different than Lactantius's. He feared lest the antipodes would cut into heaven by a sloping summit and he denied that it could except it be fixed like cats by toenails or to the globe of the earth. But Augustine does not deny that they [antipodes] can be there without danger of tottering . . . but he thinks that this region is without inhabitants because he cannot understand how the progeny of Adam or of Noah after the flood penetrated there from Armenia." Ibid., p. 37.

73. Froimond quotes from Augustine's *Epistle 18 De verbo Domini, Anti-Aristarchus* (Louvain, 1631), p. 39.

74. "For you find certain things digested in no sacred book using an orderly method for easy use and comprehension, as you have in the Aristotelian organon or Acromatica, the *Elements* of Euclid or Ptolemy's *Almagest*. But there are certain things from the disciplines scattered in passing [in Scripture] where the occasion arises or if ever longer to be dwelt on. The Holy Spirit judged it useful to speak the praises of the most wise Maker in the structure [*fabrica*] of this world and by the administration of all nature. But it is as manifest as it can be that these things are not sufficient for the fullness of theory [*doctrina*]." Froimond, *Anti-Aristarchus* (Louvain, 1631), p. 41.

75. Jacob Lansbergen, *Apologia pro commentationibus Philippi Lansbergii in Motuum Terrae Durnum et Anuum adversus Libertum Fromondum Theologum Lovaniensem et Joan. Baptistam Moirnum Doct Med. et Pariis Mathematum Professorem Regium* (Middelburg, 1633), p. 116.

76. Jacob Lansbergen, *Apologia* (Middelburg, 1633), p. 116.

77. Ibid., p. 117.

78. "That the structure of this world machine on the Copernican hypothesis is beautiful is taught by the facts themselves. Whether the position and order of the spheres or the motion of celestial bodies . . . There is nothing lacking in it, nothing redundant. . . . The Copernican hypothesis also enhances the power of God. It makes the power of God incomprehensible because not only did he make it almost infinite in number, he also created this visible world so large that although it is finite, it appears like infinity." Ibid., p. 118.

79. "Third, the goodness of God the maker shines much in the motion and position of the earth which the Copernican hypothesis attributes to it in the ecliptic. God testifies in Deut 31:11, 'like an eagle that stirs up its nest and hovers over its young.' By the earth's motion in the ecliptic, God daily presents to all men what he says specifically about the Israelites, namely, that by his powerful hand he moves the terrestrial globe from place to place. David exclaims in Psalm 8, 'what is man that you think of him and the son of man that you remember him for you made him a little lower than the angels and crowned him with glory and honor.' This incomprehensible love of God for man is discerned more clearly in the Copernican hypothesis than in the contrary." Ibid., p. 119.

80. "[Mr. Lansbergen] asks since God and nature does nothing in vain, why he has revealed to us something of so great moment. He responds that God admonishes us not only that the earth but heaven too was made especially for our sake and that even as we now possess earth, so we shall after this life possess heaven." Ibid., p. 122.

81. "The diurnal motion of the earth constantly preserves this order and the highest standard of proportion but a stationary earth violates and inverts it. For it [an earth at rest] ascribes to the highest and largest sphere the fastest and quickest motion and at the same time has all the spheres that are unequal in magnitude going around in only one diurnal revolution. This is absurd and alien to the heavens." Ibid., p. 2.

82. Ibid., p. 45.

83. "Froimond does not see that the Apostle himself expressly intends a limit," ibid., p. 44, and "even though heretics use Scripture, still dogmas of faith ought to be taken from it because they cannot be sought elsewhere." Ibid., p. 46.

84. "But, Froimond, this [imprecise value] is what Lansbergen urged and argued. If it is rough and mechanical and deviates a lot from the precise truth of the proportion, it is not geometry nor consequently ought the geometrical proportion of a circle to the diameter be taken from Scripture." Ibid., p. 44.

85. Ibid., p. 48.

86. "Theologians of the Reformed religion also hold this opinion [accommodated language] for which our Calvin alone will suffice." Ibid., p. 49.

87. Lansbergen cites Job 41, but this chapter does not have any references to thunder. Perhaps he was thinking of Job 40:9. See Lansbergen, *Apologia*, p. 51.

88. "The holy interpreters answer the calumny [of the pagans] better by showing when the occasion arose that Sacred Scripture taught nothing other than what agreed with true philosophy to a tee. For this reason also St. Augustine uses this rule in interpreting Scripture that whatever about nature can be demonstrated by true proofs, we show that they are not contrary to our Scriptures." Ibid., p. 53.

89. Libert Froimond, *Vesta sive Ant-Aristarchi Vindex adversus Iac. Lansbergium Philipp F. Medicum Middelburgensem* (Antwerp, 1634).

90. Libert Froimond, *Vesta*, p. 78.

91. "You are a perfectly odious man and a great babbler." Froimond, *Vesta*, p. 78.

92. "Why did you command me to hear the Holy Spirit in Job when you try to prove that the element of water does not exceed or transgress its limit. . . . why do you want me to listen when in your judgment nothing certain for comprehension can be understood?" Ibid.

93. "As often as Scripture speaks to us of an earth at rest and heaven in motion, the Copernicans will interpret it likewise as babbling to us in the common errors." Ibid., p. 85.

94. Ibid.

95. Among the extensive literature on Cartesiansim in the Netherlands, see the recent studies, Ernestine van der Wall, "Orthodoxy and Skepticism in the Early Dutch Enlightenment" in *Skepticism and Irreligion in the Seventeenth and Eighteenth Centuries*, Richard Popkin and Arjo Vanderjagt, eds. (Leiden: E.J. Brill, 1993) and Theo Verbeek, *Descartes and the Dutch Early Reactions to Cartesian Philosophy 1637–1650* (Carbondale: Southern Illinois University Press, 1992). Many other Protestant groups also had varying reactions to Cartesianism. On the Mennonites, see H. A. M. Snelders, "Science and Religion in the Seventeenth Century: The case of the Northern Netherlands" in C. S. Maffioli and L. C. Palm, eds., *Italian Scientists in the Low Countries in the XVIIth and XVIIIth Centuries* (Amsterdam, 1989), p. 70.

96. Scheurleer, Th. H. Lunsingh and G. H. M. Posthumus Meyjes, *Leiden University in the Seventeenth Century* (Leiden, 1975).

97. J. J. Woltjer "Introduction" in Scheurleer, Th. H. Lunsingh and G. H. M. Posthumus Meyjes, *Leiden University in the Seventeenth Century*, pp. 5, 6.

98. P. A. G. Dibon, *L'enseignement philosophique dans les universités néerlandaises à l'époque pré-cartésiens, 1575–1650* (Leiden, 1954).

99. On Cocceius's views on the Sabbath and his battle with Voet, see C. Louise Thijssen-Schoute, *Nederlands Cartesianisme* (Amsterdam, 1954), p. 31.

100. A discussion of some of Cocceius's disciples influenced by Cartesian thought can be found in C. Louise Thijssen-Schoute, "Le Cartesianisme aux Pays-Bas" in E. J. Dijksterhuis et al. *Descartes et le Cartésianisme hollandais Etudes and Documents* (Presses universitaires de France, 1950).

101. Johannes Cocceius, *Commentationes in Epistolam ad Philippenses* (1669) in J. N. Bakhuizen van den Brink, W. F. Dankbaar, W. J. Kooiman, D. Nauta and N. van der Zijp, *Documenta Reformatoria* (Kampen, 1960), pp. 419, 420.

102. Abraham Heidanus, *Consideratien, over Eenge saecken onlanghs Voorgevallen in de Universiteyt Leyden,* corrected 2nd ed. (Leiden: Arnout Doude, 1676); excerpts reprinted in *Documenta Reformatoria,* p. 440.

103. Heidanus, *Consideratien.*

104. Biographical information on Voet (Latin Voetius) can be found in *Nieuw Nederlands Biographisch Woordenboek,* vol. 7, pp. 1279–81.

105. The standard work on Regius's life and thought is M. J. A. de Vrijer, *Henricus Regius, en "Cartesiaansch" hoogleraar aan de Utrechtsche Hoogeschool* (The Hague, 1917). Regius's major work *Philosophia Naturalis* is housed at the Instituut voor geschiedenis der Natuurwetenschappen, Utrecht.

106. *Thersites heautontimorumenos hoc est Remonstrantium hyperaspistes catechesi, et liturgiae Germanicae, Gallicae & Belgicae denuo insultans, retusus idemque provocatus ad probationem Mendaciorum & Calumniaru, quae in Illustr. D.D. Ordd. & Ampliss. Magistratus Belgii, Religionem Reformatum, Ecclesias, Synodos, Pastores etc. sine ratione, sine modo effudit A Gisberto Voetio Sacrarum literarum in illudtri Gynasio Ultrajectino Professore* (Utrecht, 1635), p. 266.

107. *Thersites heautontimorumenos,* p. 266.

108. See "Bois, Jacob du" in *Nieuw Nederlandsch Biographisch Woordenboek,* vol. 4, pp. 193–94.

109. Christoph Wittich, *Consideratio theologica de stylo scripturae quem adhibet cum rebus naturalibus sermonem instituit* (1656), but apparently written before 1655.

110. The full titles of Bois's works are: *Naecktheyt Van de Cartesiaensche Philosophie, Ontbloot in een Antwoort Op een Cartesiaensch Libel, Genaemt Bewys, dat het gevoelen van die gene, die leeren der Sonnestilstandt, en des Aerdtrijcks beweging niet strydig is met Gods Woort* (Utrecht, 1655) and *Schadelickheyt Van de Cartesiaensche Philosophie, Ofte Klaer Bewijs, hoe schadelick die Philosophie is, soo in het los maecken van Godes H. Woordt, als in het invoeren van nieuwe schadelicke Leeringen. Tot antwoort Op de tweede en vermeerderde druck van Doct. Velthuysens Bewys* (Utrecht, 1656). The full titles of Velthuysen's works are: *Bewys Dat noch de Leere van der Sonne Stilstand, En*

des Aertryx Bewegingh, Noch de gronden vande Philosophie van Renatus Des Cartes strijdig sijn met Godts woort. Gestelt tegen een Tractaet van J. du Bois, Predikant tot Leyden: Genaemt Naecktheyt vande Cartesiaensche Philosophie, onbloot, etc. (Utrecht, 1656); the Latin version is entitled *Demonstratio, in qua ostenditur, neque doctrina de quiete Solis et motu Terrae neque principia Philosophiae Renati Descartes verbo Dei contraria esse, opposita Tractatui J. du Bois, Concionatoris Ludgduni Batavorum. Cui titulis Nova Philosophiae Carthesianae;* Velthuysen's second work is: *Nader Bewys. Dat noch de Leere van der Sonne Stilstant, En des Aertryx Beweging, Noch de gronden van de Philosophie van Renatus Des Cartes strydig sijn met Godts Woort. Gestelt tegen een Tractaet van J. du Bois, Predikant tot Leyde; Genaemt Schadelickheyt van de Cartesiaensche Philosophie etc.* (Utrecht, 1657). The Latin version is *Luculentior Demonstratio.* I have not had access to de Bois's works directly, but they are quoted extensively in Velthuysen's works. In any case, my main burden here is to explain the latter's views.

111. Throughout his treatise Velthuysen spoke of the views of the Reformers as a guide to his contemporary situation. "I consider it heedless of some to think their neighbor impious who hold to the same religion and the common opinion of the Reformers about the authority of Sacred Scripture since there should be liberty for each to think in this controversy what seems right. For it should be considered Papist to condemn the view of those who hold the motion of the earth" Velthuysen, *Demonstratio* in *Opera* (Rotterdam, 1680), p. 1045.

112. Velthuysen's opponents claimed that "The cause of God and the Church should be defended in this matter and it should be taught that this view is contrary to the clear texts of Scripture. It is impious to deny these texts because denying them endangers the authority of Sacred Scripture and therefore the Christian religion." Ibid., p. 1046.

113. Ibid., p. 1047.

114. "He [de Bois] rejects the distinction which I make between saying (dicere) and teaching (dicere) or making dogma (dogmatizare); he maintains that the Holy Spirit with all the words he uses teaches and intends dogma." Ibid., p. 1069.

115. Quoted by Velthuysen in ibid., p. 1069.

116. Ibid.

117. Velthuysen also compares de Bois's method to Roman Catholic methods in ibid., p. 1070.

118. "It is always true that whatever is perceived clearly and distinctly should be taken as truth. . . . when the Sacred Page accustomed to only words of diverse texts, gives contraries, it should be decided by reason where the words would be improperly accepted. An example of this we gave above cer-

tainly when Sacred Scripture speaking of God as a Spirit ascribes hands to Him." Ibid., p. 1047.

119. "But because the Holy Spirit does not so use the words of Sacred Scripture, that he wants the sense to be perceived from the words of Sacred Scripture themselves but that (as is frequently the case among men) he leaves to the human mind to draw the sense of the words from other signs, one should see whether the circumstances or the modes of speech in the texts which speak of the motion of the sun do not admit a sense of the words improper to use." Ibid., p. 1048.

120. Ibid., p. 1135.

121. "The word of God is properly its sense and not the letter (2 Pet 3:16; 1 Cor 2:14, 15; John 6:63) so no human exposition of that word can be taken as the word of God or as a rule of faith (1 Cor 7:23; 3:20). So neither can theology be a rule of faith for others as far as it is doctrine drawn from the word of God according to an exposition of men about the word of God . . . The judgment and faith of each Christian ought to be based on his own knowledge and experience founded on the word of God, not on the judgment of synods and classes." Ibid., pp. 1135, 1136.

122. *Dissertationes duae, una de usu et abusu acripturae in rebus philosophicis, altera de Cartesii sententia de quiete terrae* (1653) housed in the library of the Gemeente Universiteit of Amsterdam.

123. To my knowledge the only exposition of Wittich's views of natural philosophy and theology is Mauro Pesce, "Il Consensus Veritatis di Christoph Wittich e la distinzione tra verita scientifica e verita biblica," *Annali di storia dell'esegesi* 9 (1992): 53–76.

124. Christoph Wittich, *Consensus veritatis in Scriptura divina et infallibili revelatae cum veritate philosophica a Renato des Cartes detecta* (Nijmegen, 1659), preface, p. 5. Hereafter cited as *Consensus veritatis*.

125. *Consensus veritatis*, preface, p. 5.

126. Wittich, *Consensus veritatis*, preface, p. 8.

127. For Descartes's account of the earth being a former star and falling into the solar vortex, see *Principles of Philosophy*, Valentine Miller and Reese Miller, trans., part IV (of the Earth), no. 2, pp. 181, 182.

128. Wittich, *Consensus veritatis*, chap. XVIII, proposition xxvii, p. 185. See also pp. 182–84 for the background to Wittich's Cartesian explanation of annual motion.

129. Ibid., proposition xxviii, p. 186.

130. Ibid., p. 187.

131. "So my view is the same as Calvin's, namely, that Scripture speaks about nature according to appearances or that it uses formulas in relating the

truth to human senses and gives a complicated truth by referring to prejudices. It is surprising that the adversaries dare to distort the clear words of
Calvin." Ibid., chap. XXI, p. 215.

132. "So Scripture, when it chose to use a human style, the Holy Spirit did
not want to write with a new language but wanted to use something that was
received and well-known among men. By the same manner of speech about
things that fall under our senses, it had to speak with the customs of men and
all the more this manner of speech expresses some truth that can be recognized by all men." Ibid., p. 213.

133. "Thus Junius, Calvin and the Genevans." Ibid., p. 214.

6. COPERNICANISM AND THE BIBLE IN CATHOLIC EUROPE

1. The text of the Decree can be found in Maurice A. Finocchiaro's *The
Galileo Affair: A Documentary History* (Berkeley: University of California
Press, 1989), pp. 148–49. An English translation of Foscarini's *Lettera* can be
found in Richard J. Blackwell, *Galileo, Bellarmine and the Bible* (Notre Dame:
University of Notre Dame Press, 1991), Appendix 6, pp. 217–51.

2. So argued by Bruno Basile, "Galileo e il teologo 'copernicano' Paolo
Antonio Foscarini," *Rivista di Letteratura Italiana* 1 (1983): 63–96.

3. Vincent of Lérins, *Commonitorium*, excerpted in Conrad Kirch, S.J.,
Enchiridion Fontium Ecclesiasticae Antiquae (Barcelona: Editorial Herder,
1947), pp. 466–67.

4. Norman Tanner, *Decrees of the Ecumenical Councils* (Washington,
D.C.: Georgetown University Press, 1990), vol. 2, p. 664.

5. James M. Lattis, *Between Copernicus and Galileo: Christoph Clavius
and the Collapse of Ptolemaic Astronomy* (Chicago: University of Chicago
Press, 1994), pp. 123–25, 139–40.

6. "I profess first with all the modesty proper to a Christian and to a
member of a religious order, that everything I am about to say is reverently
submitted, now and always to the judgment of the Holy Church, prostrating
myself at the feet of the Highest Pastor." *Letter of the Foscarini on the
Pythagorean Opinion*, trans. Richard Blackwell in *Galileo, Bellarmine and the
Bible* (Notre Dame: University of Notre Dame Press, 1991), p. 226.

7. Maurice A. Finocchiaro, "The Methodological Background to Galileo's
Trial" in William A. Wallace, ed., *Reinterpreting Galileo* (Washington, D.C.:
University of America Press, 1986), pp. 241–72.

8. Ernan McMullin, "Galileo on Science and Scripture" in Peter Machamer,
ed., *The Cambridge Companion to Galileo* (Cambridge, Cambridge University
Press, 1998), pp. 217–347. For the principles, see McMullin, pp. 291–99.

9. "The exegetical positions laid out in the *Letter to the Grand Duchess* are already contained in germ in Augustine's work. Despite the claims made for it in recent Galileo scholarship, Galileo's contribution to exegesis was not especially novel." McMullin, p. 299.

10. Galileo's dependence on Augustine for his argumentation is elaborated more fully in Kenneth J. Howell's "Galileo and the History of Hermeneutics" in Jitse van der Meer, ed., *Facets of Faith and Science* (University Press of America, 1994).

11. Recently, Pietro Redondi has argued for a much closer connection between Galileo's science and his theology. See Pietro Redondi, "From Galileo to Augustine," in Peter Machmer, ed., *The Cambridge Companion to Galileo*, pp. 175–210.

12. *Castelli to Galileo*, December 14, 1613, trans. Maurice A. Finocchiaro in *The Galileo Affair*, pp. 47, 48.

13. *Galileo to Castelli* in Finocchiaro, *The Galileo Affair*, p. 49. See also Stillman Drake's translation in *Galileo at Work* (Chicago: University of Chicago Press, 1978), p. 224: "occasion to consider *again* some general things concerning the carrying of Holy Scripture into disputes about physical conclusions" (emphasis mine).

14. Drake, *Galileo at Work*, p. 224.

15. Ibid., p. 240.

16. Galileo's concern to protect theology as well as science is evident in several statements in the *Letter to Christina*. See *Le Opere di Galileo Galilei*, Antonio Favaro, ed., vol. 5, pp. 320–22; for corresponding pages in English trans., see Finocchiaro, *The Galileo Affair*, pp. 96–98.

17. *Opere*, vol. 5, p. 325, and Finocchiaro, *The Galileo Affair*, p. 325.

18. "Many dispute much about these matters which our authors wisely omitted, teaching things not useful for eternal life (*ad beatam vitam*). They devoted their attention to more valuable things giving more time to helpful matters. For what does it matter to me whether heaven encloses earth on every side like a sphere with earth placed in the middle or whether heaven partly hides the earth from above? But this is of no consequence because the faith of the Scriptures (*de fide Scripturarum*) is being dealt with . . . briefly said our authors did know what the truth was but the Spirit of God speaking through them did not want to teach men things that were not profitable for salvation." *De Genesi ad Litteram* 2, 9, para. 20, in J. P. Migne, *Patrologia Latina* (Paris: Garnier, 1878) 34, p. 270 (trans. mine). This passage is quoted by Galileo to argue that "the Holy Spirit did not want to teach us whether heaven moves or stands still, not whether its shape is spherical or like extended along a plane, nor whether the earth is located at its center or on one side." *Letter to Christina*, in Finocchairo, *The Galileo Affair*, p. 95. Galileo's list of questions is exactly those mentioned by Augustine in Book 2, chap. 9.

19. Cf. *Considerazioni circa l'opinione Copernicana* in *Opere*, vol. 5, pp. 277ff., where Galileo states that the unanimous agreement of the Fathers required by the Council of Trent applies only to "matters of faith and morals." Galileo thus believed his position to be consistent with the Council's decree.

20. "Because two truths cannot contradict one another, the task of the wise interpreter is to strive to fathom the true meaning of the sacred texts; this will undoubtedly agree with those physical conclusions of which we are already certain and sure through clear observations or necessary demonstrations." *Opere*, vol. 5, p. 320, and Finocchiaro, *The Galileo Affair*, p. 96.

21. See Moss, "Galileo's *Letter to Christina*," p. 567, for extensive documentation of references to "necessary demonstrations" and "sensate experiences" in the *Letter to Christina*. I will not enter into the philosophical dispute on whether Galileo had demonstrations in some modern sense. My only purpose is to show that he believed he had them or would have them in the very near future.

22. Galileo's first *Letter on Sunspots*, trans. in Stillman Drake, *Discoveries and Opinions of Galileo* (New York: Doubleday Anchor Books, 1957), p. 97. See also Wallace, *Galileo and His Sources*, pp. 288–91, for a discussion of the *Letter on Sunspots* with respect to Galileo's methodology.

23. "For it is true that divine authority speaks more powerfully than that which human weakness conjectures. But if they can prove by such proofs that cannot be doubted, then it is to be demonstrated that what is spoken among us as a garment is not contrary to these true reasons (*veris illis rationibus*)." *De Genesi ad Litteram* 2, 9, para. 21, in Migne, *Patrologia Latina* 34, p. 270.

24. Note similarity of Galileo's words with the first sentence quoted from Augustine, "even in regard to those propositions which are not articles of faith, the authority of the same Holy Writ should have priority over the authority of any human writings containing pure narration or even probable reasons, but no demonstrative proofs; this principle should be considered appropriate and necessary inasmuch as divine wisdom surpasses all human judgment and speculation." *Letter to Christina* in Finocchiaro, *The Galileo Affair*, p. 94.

25. The troublesome character of this section was evident early in the seventeenth century. John Wilkins, in his *A Discourse Concerning a New Planet*, rejected Galileo's literal interpretation. See *The Mathematical and Philosophical Works of the Right Reverend John Wilkins* (London, 1707–8). On Wilkins, see Hooykaas's discussion in *Rheticus's Treatise*, p. 176.

26. Augustine prefers a nonliteral interpretation, but his main argument is that even the literal is not contrary to the spherical theory. See *De Genesi ad litteram* 2, 9, para. 22, in Migne, *PL* 34, p. 271. It may be that Galileo did not quote

Augustine at this point but rather cited Augustine's *De mirabilibus sacrae scripturae* 2, 4, (in Migne, *PL* 35, pp. 2175–76) because he wanted more definite proof of the Father's belief that all the celestial spheres (including the sun) stopped. See Galileo, *Opere*, vol. 5, p. 344 (trans. Finocchiaro, *The Galileo Affair*, p. 115).

27. Bellarmine, *Letter to Foscarini* in Finocchiaro, *The Galileo Affair*, p. 68.

28. The tower argument has been the subject of extensive discussions. Feyerabend's well-known indictment of Galileo in his *Against Method* (Atlantic Highlands, N.J.: Humanities Press, 1975) has been roundly criticized by P. K. Machamer in "Feyerabend and Galileo: The Interaction of Theories and the Reinterpretation of Experience," *Stud. Hist. Phil. Sci.* 4 (1973): 1–46 and by Goosens with specific reference to the tower argument in "Galileo's Response to the Tower Argument," *Stud. Hist. Phil. Sci.* 11 (1980): 215–27. For the history behind the tower argument, see Stillman Drake, "The Tower Argument in the *Dialogue*," *Annals of Science* 45 (1988): 295–302.

29. See *Dialogue Concerning the Two Chief World Systems*, trans. Stillman Drake (Berkeley: University of California Press, 1962), p. 114f.; see Ernan McMullin's introduction in *Galileo Man of Science* (New York: Basic Books, 1967), pp. 25, 26 and Dudley Shapere, *Galileo: A Philosophical Study* (Chicago: University of Chicago Press, 1974), pp. 62–64, 99–101, for a clear presentation of Galileo's relativity principle. This principle has also been emphasized by Copernicus in *De Revolutionibus* 1.5. See Michael Heller, "Galileo's Relativity" in G. V. Coyne, M. Heller and J. Zycinski, *The Galileo Affair* (Proceedings of the Cracow Conference [Vatican City: specola vaticana, 1984]), pp. 113–24.

30. Galileo, *Dialogue Concerning the Two Chief World Systems*, trans. Stillman Drake (Berkeley: University of California Press, 1962), 2nd ed.

31. Richard J. Blackwell, *Galileo, Bellarmine and the Bible* (Notre Dame: University of Notre Dame Press, 1991), pp. 233, 235.

32. Ibid., p. 234.

33. Ibid., Appendix 6, pp. 217–251. See pp. 220, 222–223 for the text and pp. 90–94 for Blackwell's discussion.

34. Quoted in ibid., p. 102. The entire *Defensio* is translated in Appendix 7B.

35. *Bellarmine to Foscarini* in Favaro, *Opere*, vol. 12, p. 160. See the English trans. in Finocchiaro, *The Galileo Affair*, pp. 67–69.

36. Bellarmine cites Ecclesiastes 1 about the sun's motion and concludes, "it is not likely that he was affirming something that was contrary to truth already demonstrated or capable of being demonstrated." Ibid. See the English translation in ibid., p. 68.

37. Bellarmine's views on astronomy are discussed in greater detail in Ugo Baldini, "L'astronomia del Cardinale Bellarmino" in P. Galluzzi, ed., *Novita Celesti e Crisi del Sapere* (Florence: Giunti Barbera, 1984), pp. 293–305.

38. This is Finocchiaro's translation of the Latin *ex suppositione*, p. 67. Others generally translate it "hypothetically" (see note 45, p. 333, in Finocchiaro, *The Galileo Affair*).

39. Ingoli's treatise can be found in A. Favaro, ed., *Le opere di Galileo Galilei*. Galileo's response to Ingoli is in vol. 6 of Galileo's *Opere*, pp. 509–561. An English translation of Galileo's reply is in Finocchiaro, *The Galileo Affair*, pp. 154–97.

40. Francesco Ingoli, *De situ et quiete Terrae contra Copernici systema Disputatio*, in Antonio Favaro, ed., *Le Opere di Galileo Galilei*, vol. 5, pp. 398–411.

41. Ibid.

42. See Thomas Campanella, O.P., *A Defense of Galileo, the Mathematician from Florence*, English translation, introduction, and notes by Richard J. Blackwell (Notre Dame: University of Notre Dame Press, 1994). This supercedes the flawed translation of Grant McColley ("The Defense of Galileo," *Smith College Studies in History* 22 [1937]: 3–4). A good description of the background and contents of the document can be found in Blackwell's Introduction, pp. 1–34, and in Bernardino M. Bonansea, "Campanella's Defense of Galileo" in William A. Wallace, ed., *Reinterpreting Galileo Studies in Philosophy and the History of Philosophy*, vol. 15.

43. For the date of composition, see Blackwell's discussion, pp. 19–24, and the more thorough treatment in Bonansea's discussion, pp. 211–13 (see previous note).

44. See Thomas Campanella, O.P., *A Defense of Galileo*, under chapter 2, "Arguments for Galileo," in Blackwell, p. 9. Campanella's conviction partially derived from his reading of *Sidereus Nuncius* in 1611. See Bonansea's discussion, p. 213.

45. See Thomas Campanella, O.P., *A Defense of Galileo*, under chapter 4, "Replies to the Arguments Against Galileo Stated in Chapter 1," in Blackwell, pp. 109, 110. I have altered Blackwell's generally good translation somewhat to bring out the force of Campanella's conclusion. *Violenter* should not be translated "arbitrarily" because biblical interpretation had long spoken of doing violence to a text by distorting its clear meaning. Campanella is using a standard term in hermeneutics. I have also translated *obsequens* as "yielding to" rather than "adjusted to" because it is a strong term that fits Campanella's emphasis on the Scriptures achieving a victory over the philosophers. Most importantly, I am puzzled by Blackwell's omission of the phrase *per sensatas experientias*. Campanella's Thomism would certainly incline him to place a lot of stock in sensible proofs of a scientific proposition. He believed that empirical science had vindicated the Bible against the philosophers.

46. Campanella, *A Defense of Galileo*, p. 89.

47. "Though the Holy Council speaks on matters of morals and faith, yet it cannot be denied that the interpretation which is contrary to the consensus of the Fathers would displeasure those holy Fathers (of the Council)." Francesco Ingoli, *De Situ et quiete*.

CONCLUSION: INTERPRETING THE HISTORY OF EARLY
MODERN COSMOLOGY AND THE BIBLE

1. Hugh of St. Victor, *Didascalion*, book 7, chap. 4.

2. A comprehensive discussion of medieval cosmology is Edward Grant, *Planets, Stars and Orbs: The Medieval Cosmos, 1200–1687* (Cambridge: Cambridge University Press, 1994). See pp. 36–39.

3. See Nicholas Jardine, "Skepticism in Renaissance Astronomy: A Preliminary Study," in *Skepticism from the Renaissance to the Enlightenment*, Richard Popkin and Charles Schmitt, eds. (Wiesbaden: Otto Harrasowitz, 1987).

4. This improves on Duhem's famous distinction between realists and instrumentalists. See Nicholas Jardine, "The Forging of Modern Realism: Clavius and Kepler against the Sceptics," *Studies in the History and Philosophy of Science* 10 (1979): 141–74; idem, *The Birth of History and Philosophy of Science: Kepler's Defense of Tycho against Ursus, with Essays on Its Provenance and Significance* (Cambridge: Cambridge University Press, 1984).

5. Aurelius Augustinus, *De Doctrina Christiana, Corpus Christiana*, Series Latina, vol. 32, part IV, 1 (Turnholt: Brepols, 1962).

6. For a discussion of Ambrose's *Hexaemeron*, see Maria Grazia Mara, "Ambrose of Milan," in Angelo de Beradino, *Patrology*, vol. 4 (Westminster, Md.: Christian Classics, 1991), pp. 153, 154, and Frank E. Robbins, *The Hexaemeral Literature*, Ph.D. diss., University of Chicago, 1912, pp. 58–59.

7. Augustine warns against confusing questions of empirical knowledge with matters of faith in *Confessions*, Book 5, chap. 5. See Henry Chadwick's trans. (Oxford University Press, 1991), pp. 76, 77, and in *The Literal Meaning of Genesis*, Book 1, chap. 19. See the translation by John Hammond Taylor in the Ancient Christian Writers Series (New York: Newman Press, 1982), pp. 42–43.

8. Amos Funkenstein, *Theology and the Scientific Imagination from the Middle Ages to the 17th Century* (Princeton, 1986), pp. 215–19.

9. The main work of Abraham Ibn Ezra was *Perush ha tora* (interpretation of the Torah). See ibid. for references, pp. 215ff.

10. Bruce Moran, "Christoph Rothmann, the Copernican Theory, and Institutional and Technical Influences on the Criticism of Aristotelian cosmology," *Sixteenth Century Journal* 13 (1982): 85–103.

11. *Rothmann to Brahe*, October 13, 1588, *TBOO*, 6, p. 149. See the discussion in chap. 3.

12. Recently, Jean Dietz Moss has treated Kepler's interpretation, but only his accommodation strategy, in the *Astronomia Nova*. See *Novelties in the Heavens* (Chicago: University of Chicago Press, 1993), pp. 132–35.

Bibliography

Aiton, Eric. "Celestial Spheres and Circles." *History of Science* 19 (1981): 75–113.

Alquie, Ferdinand. *Le rapport de la science et de la religion selon Descartes, Malebranche et Spinoze*. Geneva: Droz, 1982.

Apt, A. J. "The Reception of Kepler's Astronomy in England, 1609–1650." Ph.D. thesis, Oxford University, 1982.

Ariew, Roger. "The Phases of Venus before 1610." *Studies in the History and Philosophy of Science* 18 (1987): 81–92.

Ashworth, William B. "Catholicism and Early Modern Science." In *God and Nature: Historical Essays on the Encounter between Christianity and Science*, ed. David C. Lindberg and Ronald L. Numbers. Berkeley: University of California Press, 1986.

———. *Jesuit Science in the Age of Galileo: An Exhibition of Rare Books from the History of Science Collection*. Kansas City: Linda Hall Library, 1986.

Auger, Leon. "Les idéés de Roberval sur le système du monde." *Revue d'histoire des sciences* 10 (1957): 226–34.

Augustine, Aurelius. *De Doctrina Christiana*. Corpus Christiana, Series Latina, vol. 32, pt. IV, 1. Turnhout: Brepols, 1962.

———. *Confessions*, trans. Henry Chadwick. Oxford: Oxford University Press, 1991.

———. *The Literal Meaning of Genesis*, ed. and trans. John Hammond Taylor. Ancient Christian Writers Series. New York: Newman Press, 1982.

Ayers, Robert H. "The View of Medieval Biblical Exegesis in Calvin's *Institutes*." *Perspectives in Religious Studies* 7 (1980): 188–93.

Bakhuizen van den Brink, J. N., W. F. Dankbaar, W. J. Kooiman, D. Nauta, and N. van der Zijp. *Documenta Reformatoria*. Kampen: J. H. Kok, 1960.

Baldini, Ugo. "La Nova del 1604: i Matematici e Filosofi del Collegio Romano: Note su un Testo Inedito." *Annali dell'Istitutio e Museo di Storia di Scienza di Firenze* 4 (2) (1981): 63–78.

————. *Christoph Clavius and the Scientific Scene in Rome.* Vatican City: Pontificia Academia Scientiarum, Specola Vaticana, 1983.

————. "L'astronomia del Cardinale Bellarmino." In *Novità Celesti e Crisi del Sapere,* ed. P. Galluzzi, 293–305. Florence: Giunti Barbera, 1984.

————. "Verso una definizione storica della 'filosofia' del galileismo: Gli epistolari come strumento interpretativo." *Rivista di Storia della Filosophia* 42 (1987): 213–35.

Baldini, Ugo, and G. V. Coyne, eds. and trans. *The Louvain Lectures of Bellarmine and the Autograph Copy of His 1616 Declaration to Galileo.* Studi Galileini, vol. 1, no. 2. Vatican City: Vatican Observatory Publications, 1984.

Balestra, Dominic J. "The Case of Galileo and the New Priesthood: An Inquiry into the Grounds of Science's Claim to Our Intellectual Allegiance." *Proceedings of the American Catholic Philosophical Association* 59 (1985): 319–30.

Bangert, William V. *A Bibliographical Essay on the History of the Society of Jesus: Books in English.* St. Louis: Institute of Jesuit Sources, 1976.

Barcaro, Umberto. "Le proprietà del cerchio nei *Discorsi* di Galileo." *Physis* 24 (1982): 5–16.

Barker, Peter. "Jean Peña and the Stoic Physics in the Sixteenth Century." *The Southern Journal of Philosophy* 23 (1985): 93–107.

————. "Stoic Contributions to Early Modern Science." In *Atoms, Pneuma, and Tranquility: Epicurean and Stoic Themes in European Thought,* ed. Margaret J. Osler. Cambridge: Cambridge University Press, 1991.

Barker, Peter, and B. R. Goldstein. "Distance and Velocity in Kepler's Astronomy." *Annals of Science* 51 (1994): 59–73.

Barocius, Franciscus. *Cosmographia.* Venice, 1585.

Barone, Francesco. "Diego de Zuñiga e Galileo Galilei: Astronomia eliostatica ed esegesi biblica." *Critica Storia* 19 (1982): 319–34.

Basile, Bruno. "Galileo e il teologo 'copernicano' Paolo Antonio Foscarini." *Rivista della Letteratura Italiana* 1 (1983): 63–96.

Bechler, Zev. "The Essence and Soul of 17th-century Scientific Revolution." *Science in Context* 1 (1987): 87–101.

Becker, George. "The Fallacy of the Received Word: A Reexamination of Merton's Pietism-Science Thesis." *American Journal of Sociology* 91 (1986): 1203–18.

Bergman, Jerry. "The Establishment of a Heliocentric View of the Universe." *Journal of the History of Astronomy* 33 (1981): 225–30.

Berkel, K. van. "Universiteit en natuurwetenschap in de 17e eeuw, in het bijzonder in de Republiek." In *Natuurwetenschappen van Renaissance tot Darwin*, ed. H. A. M. Snelders and K. van Berkel, 107–30. The Hague: Martinus Nijhoff, 1981.

———. *Issac Beeckman (1588–1637), en de mechanisering van het wereldbeeld.* Amsterdam: Rodopi, 1983.

———. "Descartes in debat met Voetius. De mislukte introductie van het cartesianisme aan de Utrechtse Universiteit (1639–1645)." *Tijdschrift voor de Geschiedenis der Geneeskunde, Natuurwetenschappen, Wiskunde en Techniek* 7 (1984): 4–18.

———. *In het voetspoor van Stevin. Geschiedenis van de natuurwetenschap in Nederland, 1580–1940.* Meppel: Boom, 1985.

Bernardini, Amalia. *Antonio Arnauld: Racionalismo cartesiano y teologia: Descartes-Malebranche-Leibniz.* Costa Rica: Editorial Universidad Estatal a Distancia, 1984.

Biagioli, Mario. "Galileo the Emblem Maker." *Isis* 81 (1990): 230–58.

———. *Galileo Courtier.* Chicago: University of Chicago Press, 1993.

Bilinski, Bronislaw. *Il periodo padovano di Niccolo Copernico*, ed. A. Poppi. Scienza e filosofia all'Università di Padova nel Quattrocento. Padua: LINT, 1983.

Birkenmajer, A. "Le commentaire inédit d'Erasme Reinhold sur le 'De Revolutionibus' de Nicolas Copernic." *La science au seizième siècle* (1960): 171–77.

Blackwell, Richard J. *Galileo, Bellarmine and the Bible.* Notre Dame: University of Notre Dame Press, 1991.

Blair, Ann. "Tycho Brahe's Critique of Copernicus and the Copernican System." *Journal of the History of Ideas* 51 (1990): 355–77.

Blumenberg, Hans. *The Genesis of the Copernican World.* Cambridge, Mass.: MIT Press, 1987.

Boas, Marie. *The Scientific Renaissance.* New York: Harper, 1966.

Bogazzi, Riccardo. "Il Kosmotheoros di Christiaan Huygens." *Physis* 19 (1977): 87–109.

Bonansea, Bernardino M. "Campanella's Defense of Galileo." In *Reinterpreting Galileo Studies in Philosophy and the History of Philosophy*, vol. 15, ed. William A. Wallace. Washington, D.C.: Catholic University of America Press, 1986.

Bornkamm, Heinrich. "Kopernikus im Urteil der Reformation." *Archiv für Reformationsgeschichte* 40 (1973).

Bosmans, H. "Philippe van Lansberge, de Gand, 1561–1632." *Mathésis: recueil mathématique* 42 (1928): 5–10.

Bowen, William R., and Konrad Eisenbichler. *Published Books 1499 to 1700 on Science, Medicine, and Natural History at the Centre for Reformation and*

Renaissance Studies. Victoria College, University of Toronto: Centre for Reformation and Renaissance Studies, 1985.

Brackenridge, J. Bruce. "Kepler, Elliptical Orbits and Celestial Circularity: A Study in the Persistence of Metaphysical Commitment." *Annals of Science* 39 (1982): 117–43.

Brahe, Tycho. *Tychonis Brahe Dani Opera Omnia*, 15 vols., ed. J. L. E. Dreyer. Copenhagen: Nielsen & Lydiche, 1913.

———. "Astrologia." In John Christianson, "Tycho Brahe's Cosmology from the 'Astrologia' of 1591." *Isis* 59 (1968): 312–18.

Bretagnon, P., J. L. Simon, and J. Laskar. "Presentation of New Solar and Planetary Tables of Interest for Historical Calculations." *Journal of the History of Astronomy* 17 (1986): 39–50.

Bretschneider, Carl. *Corpus Reformatorum.* 1834.

Broad, William J. "A Bibliophile's Quest for Copernicus." *Science* 218 (1982): 661–64.

Brodrick, James, S.J. *The Life and Work of Blessed Robert Francis Cardinal Bellarmine, S.J., 1542–1621.* 2 vols. London: Burns & Oates, 1928.

Brooke, John H. *Science and Religion: Some Historical Perspectives.* Cambridge: Cambridge University Press, 1991.

Brown, Hanbury. *The Wisdom of Science: Its Relevance to Culture and Religion.* Cambridge: Cambridge University Press, 1986.

Buchwald, Betsey B. P. "The Astronomy of Albertus Magnus." Ph.D. thesis, University of Toronto, 1983.

Burke, John G., ed. *The Uses of Science in the Age of Newton.* Berkeley: University of California Press, 1983.

Burmeister, Karl H. *Georg Joachim Rheticus, 1514–1574, Eine Bio-Bibliographie.* 3 vols. Wiesbaden: Guido Pressler Verlag, 1966.

Butterfield, Herbert. *The Origins of Modern Science.* New York: Collins, 1962.

Calvin, John. *Institutes of the Christian Religion.* 1555 edition.

———. *Commentaries*, ed. Joseph Haroutunian and Louise Smith. London: Library of Christian Classics, 23).

———. *Commentary on Psalms.* Grand Rapids: Eerdmans, 1963.

———. *Concerning Scandals*, trans. John W. Fraser. Grand Rapids: Eerdmans, 1978.

Campadelli, Luigi. "Riccioli, Giambattista." In *Dictionary of Scientific Biography* 11:411–12.

Campanella, Thomas. *A Defense of Galileo*, ed. and trans. Richard Blackwell. Notre Dame: University of Notre Dame Press, 1994.

Carugo, Adriano, and Alistair C. Crombie. "The Jesuits and Galileo's Ideas of Science and of Nature." *Annali dell'Istituto e Museo di Storia de Scienza di Firenze* 8 (2) (1983): 3–68.

Caspar, Max. *Kepler*. New York: Abelard-Schumann, 1959.

Chalmers, Alan. "Planetary Distances in Copernican Theory." *British Journal for the Philosophy of Science* 32 (1981): 374–75.

Chalmers, Alan, and Richard Nicholas. "Galileo on the Dissipative Effect of a Rotation of the Earth." *Studies in the History and Philosophy of Science* 14 (1983): 315–40.

Chant, Colin, and John Fauvel, eds. *Darwin to Einstein: Historical Studies on Science and Belief*. London: Longman, 1980.

Christianson, J. R. "Copernicus and the Lutherans." *Sixteenth Century Journal* 4 (October 1973): 1–10.

———. *On Tycho's Island*. Cambridge: Cambridge University Press, 2000.

Clarke, Desmond M. *Descartes' Philosophy of Science*. Manchester: Manchester University Press, 1982.

Clavelin, Maurice. *Conceptual and Technical Aspects of the Galilean Geometrization of the Motion of Heavenly Bodies. Nature Mathematized: Historical and Philosophical Case Studies in Classical Modern Natural Philosophy*, ed. W. R. Shea. Dordrecht: Reidel, 1983.

———. "Histoire des sciences et theorie du raisonnement: A propos d'un livre recent sur Galilee et l'art du raisonnement." *Revue d'Histoire des Sciences* 35 (1982): 331–40.

Clericuzo, Antonio. "From Van Helmont to Boyle." *The British Journal for the History of Science* 26 (1993): 303–34.

Cohen, Floris. "Open and Wide, Yet Without Height and Depth." *Tractrix* 2 (1990): 159–65.

———. *The Scientific Revolution: A Historiographical Inquiry*. Chicago: University of Chicago Press, 1994.

Cohen, I. Bernard. "The Origins of the Concept of a Copernican Revolution." *Organon* 14 (1978): 15–25.

———. "Perfect Numbers in the Copernican System: Rheticus and Huygens." In *Science and History: Studies in Honor of Edward Rosen*. Studia Copernicana, 16. Wroclaw: Ossolineum, 1978.

———. "The Influence of Theoretical Perspective on the Interpretation of Sense Data: Tycho Brahe and the New Star of 1572, and Galileo and the Mountains on the Moon." *Annali dell'Istitutio e Museo di Storia di Scienza di Firenze* 5 (1) (1980): 3–14.

———. *Revolution in Science*. Cambridge: Harvard University Press, 1985.

———, ed. *Puritanism and the Rise of Modern Science*. New Brunswick: Rutgers University Press, 1990.

Colloquia Copernicana, III. *Astronomy of Copernicus and Its Background. Proceedings of the Joint Symposium of the IAU and IUHPS, co-sponsored by the IAHS*. Studia Copernicana, 13. Wroclaw: Ossolineum, 1975.

Colloquia Copernicana, IV. *Conférences des symposia: L'audience de la theorie héliocentrique, Copernic et le développement des sciences exactes et sciences humaines.* Studia Copernicana, 14. Wroclaw: Ossolineum, 1975.

Conant, James. "The Advancement of Learning during the Puritan Commonwealth." *Proceedings of the Massachusetts Historical Society* 66 (1942): 3–31.

Copenhaver, Brian P. "Jewish Theologies of Space in the Scientific Revolution: Henry More, Joseph Raphson, Isaac Newton, and Their Predecessors." *Annals of Science* 37 (1980): 489–548.

Copernicus, Nicolaus. *Minor works*, ed. F. Pawel Czartoryski. London: Macmillan, 1985.

————. *De Revolutionibus.* In *Great Books of the Western World*, ed. Robert Hutchins. Chicago: Encyclopaedia Britannica, 1939.

Costabel, Pierre. "L'atomisme, face cachée de la condamnation de Galiléé?" *Vie de Science* 4 (1987): 349–65.

Courtenay, William J. *Covenant and Causality in Medieval Thought: Studies in Philosophy, Theology, and Economic Practice.* London: Variorum Reprints, 1984.

Cramer, Jan A. *Abraham Heidanus en zijn Cartesianisme.* Utrecht: J. van Druten, 1889.

Crowe, Michael J. *Theories of the World from Antiquity to the Copernican Revolution.* New York: Dover Books, 1990.

Cynarski, Stanislaw. *Reception of the Copernican Theory in Poland in the 17th and 18th Centuries.* Zeszyty Naukowe Uniwersytetu Jagiellonskiego, 328. Prace historyczne, zeszyt 44. Copernicana Cracoviensia 5. Cracow: Jagellonian University 1973).

Davis, Edward B. "Creation, Contingency, and Early Modern Science: The Impact of Voluntaristic Theology on 17th-Century Natural Philosophy." Ph.D. thesis, Indiana University, 1985.

Dear, Peter. "Jesuit Mathematical Science and the Reconstituion of Experience in the Early 17th Century." *Studies in the History and Philosophy of Science* 18 (1987): 133–75.

————. *Mersenne and the Learning of the Schools.* Ithaca: Cornell University Press, 1988.

de Bois, Jacob. *Naecktheyt Van de Cartesiaensche Philosophie, Ontbloot in een Antwoort Op een Cartesiaensch Libel, Genaemt Bewys, dat het gevoelen van die gene, die leeren der Sonnestilstandt, en des Aerdtrijcks beweging niet strydig is met Gods Woort.* Utrecht, 1655.

————. *Schadelickheyt Van de Cartesiaensche Philosophie, Ofte Klaer Bewijs, hoe schadelick die Philosophie is, soo in het los maecken van Godes H. Woordt, als in het invoeren van nieuwe schadelicke Leeringen. Tot antwoort Op de tweede en vermeerderde druk van Doct. Velthuysens Bewys.* Utrecht, 1656.

Debus, Allen G. "Mathematics and Nature in the Chemical Texts of the Renaissance." *Ambix* 15 (1968): 1–28.

———. *The Chemical Philosophy, Paracelsian Science, and Medicine in the Sixteenth and Seventeenth Centuries.* 2 vols. New York: Science History Publications, 1977.

———. "Key to Two Worlds: Robert Fludd's Weather-Glass." *Annal dell'Istitutio e Museo di Storia di Scienza di Firenze* 7 (1982): 109–43.

———. *Science and History: A Chemist's Appraisal.* Lectures at the University of Coimbra, 1983.

Deissman, Adolf. *Schriftauthorität bei Johannes Kepler.* Marburg: Elwert, 1894.

Descartes, René. *Principles of Philosophy,* ed. and trans. Valentine Miller and Reese Miller. Dordrecht: Reidel, 1983.

Diamond, Norman. "The Coperican Revolution: Social Foundations of Conceptualization in Science." In *Science as Politics,* ed. Les Levidow, 7–17. London: Free Association Books, 1986.

Dijksterhuis, E. J. *Simon Stevin.* The Hague: Martinus Nijhoff, 1943.

Dillenberger, John. *Protestant Thought and Natural Science: A Historical Interpretation.* New York: 1960.

———, ed. *John Calvin: Selections from His Writings.* Missoula, Mont.: Scholars Press, 1975.

Dobbs, Betty Jo. *The Hunting of the Green Lion: The Foundations of Newton's Alchemy.* Cambridge: Cambridge University Press, 1975.

———. "Newton's Commentary on the Emerald Tablet of Hermes Trismegistus: Its Scientific and Theological Significance." In *Hermeticism and the Renaissance: Intellectual History and the Occult in Early Modern Europe,* ed. Ingrid Merkel and Allen G. Debus. Washington: Folger Shakespeare Library, 1988.

———. *The Janus Face of Genius.* Cambridge: Cambridge University Press, 1991.

Dobrzycki, Jerzy. *The Reception of Copernicus' Heliocentric Theory.* Dordrecht: D. Reidel, 1972.

Doebel, Gunter. *Johannes Kepler: Er veranderte das Weltbild.* Verlag Styria, 1983.

Donahue, William H. *The Dissolution of the Celestial Spheres, 1595–1650.* New York: Arno Press, 1981.

———. "Kepler's Cosmology." In *Encyclopedia of Cosmology,* ed. Norriss Hetherington. Garland Press, 1993.

———. "Kepler's Invention of the Second Planetary Law." *The British Journal for the History of Science* 27 (1994): 89–102.

Donnelly, John Patrick. "The Jesuit College at Padua: Growth, Suppression, Attempts at Restoration (1552–1606)." *Archive for the History of the Society of Jesus* 51 (1982): 47–79.

Drake, Stillman, trans. *Letter on Sunspots in Discoveries and Opinions of Galileo.* New York: Doubleday Anchor Books, 1957.

———. *Dialogue Concerning Two Chief World Systems.* Berkeley: University of California Press, 1967.

———. *Galileo at Work.* Chicago: University of Chicago Press, 1978.

———. "Ptolemy, Galileo, and Scientific Method." *Studies in the History and Philosophy of Science* 9 (1978): 99–115.

———. *Galileo.* Oxford: Oxford University Press, 1980.

———. *Telescopes, Tides, and Tactics: A Galilean Dialogue about the Starry Messenger and Systems of the World.* Chicago: Univeristy of Chicago Press, 1983.

———. "Reexamining Galileo's *Dialogue.*" In *Reinterpreting Galileo Studies in the History and Philosophy of Science,* ed. William A. Wallace. Washington, D.C.: Catholic University of America Press, 1986.

———. "Galileo's Pre-Paduan Writings: Years, Sources, Motivations." *Studies in the History and Philosophy of Science* 17 (1986): 429–48.

———. "Galileo's Steps to Full Copernicanism and Back." *Studies in the History and Philosophy of Science* 18 (1987): 93–105.

———. "The Tower Argument in the *Dialogue.*" *Annals of Science* 45 (1988): 295–302.

Draper, John Henry. *A History of the Conflict Between Religion and Science.* New York, 1874.

Dreyer, J. L. E. *Tycho Brahe: A Picture of Scientific Life and Work in the Sixteenth Century.* New York: Dover reprint, 1963.

———. *History of Astronomy from Thales to Kepler.* New York: Dover, 1953, 2d ed.

Duhem, Pierre. *To Save the Phenomena: An Essay in the Idea of Physical Theory from Plato to Galileo.* Chicago, 1969.

———. *The Aim and Structure of Physical Theory.* Princeton: Princeton University Press, 1912.

Duker, A. C. *Gisbertus Voetius.* 2 vols. Leiden: Groen, 1989 reprint.

Eastwood, Bruce S. "Kepler as Historian of Science: Precursors of Copernican Heliocentrism according to *De revolutionibus.*" I, 10. *Proceedings of the American Philosophical Association* 126 (1982): 367–94.

Ernst, Germana. "Aspetti dell'astrologia e della profezia in Galileo e Campanella." In *Novita Celesti e Crisi del Sapere,* ed. Paolo Galluzzi, 255–66. Florence: Giunti Barbera, 1984.

Fabris, Rinaldo. *Galileo Galilei e gli orientamenti esegetici del suo tempo.* Vatican City: Pontificiae Academiae Scientiarum, 1986.

Fantoli, Annibale. *Galileo for Copernicanism and for the Church.* Vatican: Vatican Observatory Foundation, 1994.

Favaro, Antonio, ed. *Le Opere di Galileo Galilei.* 20 vols. Florence, Giunti Barbera, 1890–1909.

Feldhay, Rivka. "Knowledge and Salvation in Jesuit Culture." *Science in Context* 1 (1986): 195–213.

Field, J. V. "Kepler's Star Polyhedra." *Vistas in Astronomy* 23 (1979): 109–41.

———. *Kepler's Geometrical Cosmology.* Chicago: University of Chicago Press, 1988.

Finocchiaro, Maurice A. "The Methodological Background to Galileo's Trial." In *Reinterpreting Galileo,* ed. William A. Wallace, 241–72. Washington, D.C.: University of America Press, 1986.

———. *The Galileo Affair: A Documentary History.* Berkeley: University of California Press, 1989.

Fletcher, John M. "Change and Resistance to Change: A Consideration of the Development of English and German Universities during the 16th Century." *History of Universities* 1 (1981): 1–36.

Folsing, Albrecht. *Galileo Galilei: Prozess ohne Ende Eine Biographie.* Munich: Piper, 1983.

Freudenthal, Gad. "Theory of Matter and Cosmology in William Gilbert's *De Magnete.*" *Isis* 74 (1983): 22–37.

Fromoind, Libert. *Anti-Aristarchus sive Orbis Terrae Immobilis liber unus.* Louvain, 1631.

———. *Vesta sive Anti-Aristarchi Vindex.* Louvain, 1634.

Funkenstein, Amos. *Theology and the Scientific Imagination from the Middle Ages to the 17th Century.* Princeton: Princeton University Press, 1986.

———. "The Body of God in 17th-Century Theology and Science." In *Millenarianism and Messianism in English Literature and Thought, 1650–1800,* ed. Richard H. Popkin. Leiden: Brill, 1988.

Galluzzi, Paolo. "Galileo contro Copernico: II dibattito sulla prova 'galileiana' di G. B. Riccioli contro il moto della terre alla luce di nuovi documenti." *Annali dell'Istituto e Museo di Storia di Scienza di Firenze* 2 (2) (1977): 87–148.

———, ed. *Novita Celesti e Crisi del Sapere.* Florence: Giunti Barbera, 1984.

Gamble, Richard C. "*Brevitas et Facilitas*: Toward an Understanding of Calvin's Hermeneutic." *Westminster Theological Journal* 47 (1985): 1–17.

———. "Exposition and Method in Calvin." *Westminster Theological Journal* 49 (1987): 153–65.

Gardner, Michael R. *Realism and Instrumentalism in pre-Newtonian Astronomy.* Minneapolis: University of Minnesota Press, 1983.

Garin, Eugenio. "Il Caso Galileo nella Storia della Cultura Moderna." *Annali dell'Istituto e Museo di Storia di Scienza di Firenze* 8 (1) (1983): 3–17.

Gascoigne, John. "A Reappraisal of the Role of Universities in the Scientific Revolution." In *Reappraisals of the Scientific Revolution,* ed. Lindberg and Westman, 207–60. Cambridge: Cambridge University Press, 1990.

Gerrish, B. A. "The Reformation and the Rise of Modern Science." In *The Impact of the Church upon Its Culture,* ed. Jerald C. Brauer. Chicago: University of Chicago Press, 1968.

Geyl, Pieter. *The Netherlands in the Seventeenth Century (1609–1648).* London: Cassell, 1961.

Giard, Luce. "Histoire de l'université et histoire du savoir: Padoue." *Révue Synthèse* 104 (1983): 139–69.

Gingerich, Owen. "From Copernicus to Kepler: Heliocentrism as Model and as Reality." *Proceedings of the American Philosophical Society* 117 (1973): 516–20.

———. "'Crisis' versus Aesthetic in the Copernican Revolution." *Vistas in Astronomy* 17 (1975): 85–95.

———. "The 1582 *Theorica Orbium* of Hieronymus Vulparius." *Journal of the History of Astronomy* 8 (1977): 38–43.

———. "Early Copernican Ephemerides." In *Science History: Studies in Honor of Edward Rosen,* 403–17. Studia Copernicana, 16. Wroclaw: Ossolineum, 1978.

———. "The Search for a Plenum Universe." In *The Eye of Heaven: Ptolemy, Copernicus, Kepler,* ed. A. Beer and K. Strand. New York: American Institute of Physics, 1993.

———. "The Great Copernicus Chase." *American Scholar* 49 (1979): 81–88.

———. "The Censorship of Copernicus' *De Revolutionibus.*" *Annali dell'Istituto e Museo di Storia di Scienza di Firenze* 6 (2) (1981): 45–61.

———. "The Galileo Affair." *Scientific American* 247 (2) (1982): 132–43.

———. *The Eye of Heaven: Ptolemy, Copernicus, Kepler,* ed. A. Beer and K. Strand. New York: American Institute of Physics, 1993.

Gingerich, Owen, and Barbara Welther. *Planetary, Lunar, and Solar Postions; New and Full Moons, a.d. 1650–1805.* Philadelphia: American Philosophical Society, 1983.

Gingerich, Owen, and Robert S. Westman. *The Wittich Connection: Conflict and Priority in Late Sixteenth-Century Cosmology.* Philadelphia: Transactions of the American Philosophical Society, 1988, vol. 78, part 7.

Gingerich, Owen, and James R. Voelkel. "Tycho Brahe's Copernican Campaign." *Journal of the History of Astronomy* 29 (1998): 1–34.

Goldman, Steven Louis. "On the Interpretation of Symbols and the Christian Origins of Modern Science." *Journal of Religion* 62 (1982): 1–20.

Goldstein, Bernard R. "The Status of Models in Ancient and Medieval Astronomy." *Centaurus* 24 (1980): 132–47.

———. *Theory and Observation in Ancient and Medieval Astronomy.* London: Variorum, 1985.

Goldstein, Bernard R., and Peter Barker. "The Role of Rothmann in the Dissolution of the Celestial Spheres." *Historical and Philosophic Studies.* Princeton: Princeton University Press, 1992.

Goosens, William. "Galileo's Response to the Tower Argument." *Studies in the History and Philosophy of Science* 11 (1980): 215–27.

Grabow, Rudolf. *Simon Stevin.* Leipzig: Tuebner, 1985.

Grafton, Anthony. "Humanism and Science in Rudolphine Prague: Kepler in Context." In *Defenders of the Text: Traditions of Scholarship in an Age of Science, 1450–1800.* Cambridge, Mass.: Harvard University Press, 1991.

———. "Kepler as a Reader." *Journal of the History of Ideas* 53 (4): 561–72.

Grant, Edward. "Late Medieval Thought, Copernicus, and the Scientific Revolution." *Journal of the History of Ideas* 23 (1963): 197–220.

———. "Cosmology." In *Science in the Middle Ages,* ed. David Lindberg, 265–302. Chicago: Chicago University Press, 1978.

———. "Celestial Matter: A Medieval and Galilean Cosmological Problem." *Journal of Medieval & Renaissance Studies* 13 (1983): 157–86.

———. *In Defense of the Earth's Centrality and Immobility: Scholastic Reaction to Copernicanism in the Seventeenth Century.* Philadelphia: American Philosophical Society, 1984.

———. "Were There Significant Differences between Medieval and Early Modern Scholastic Natural Philosophy?" *Nous* 18 (1984): 5–14.

———. "Celestial Orbs in the Latin Middle Ages." *Isis* 78 (1987): 153–73.

———. *Planets, Stars and Orbs: The Medieval Cosmos, 1200–1687.* Cambridge: Cambridge University Press, 1994.

Groot, Aart de. *Inaugurele rede over Godzaligheid te verbinden met de wetenschap van Gisbertus Voetius.* Kampen: J. H. Kok, 1978.

Guerlac, Henry. "Theological Voluntarism and Biological Analogies in Newton's Physical Thought." *Journal of the History of Ideas* 44 (1983): 219–29.

Haffenreffer, Matthias. *Loci Theologici,* 4th ed., 1609.

Hammer, William. "Melanchthon, Inspirer of the Study of Astronomy; With a Translation of His Oration in Praise of Astronomy (*De Orione,* 1553)." *Popular Astronomy* 59 (1951): 308–19.

Hannaway, Owen. *The Chemists and the Word: The Didactic Origins of Chemistry.* Baltimore: The Johns Hopkins University Press, 1972.

————. "Laboratory Design and the Aim of Science: Andreas Libavius and Tycho Brahe." *Isis* 77 (1986): 585–610.

Hanson, R. P. C. *Allegory and Event.* Richmond: John Knox Press, 1959.

Harris, Stephen J. "Transposing the Merton Thesis: Apostolic Spirituality and the Establishment of the Jesuit Scientific Tradition." *Science in Context* 3 (1989): 29–65.

Hartfelder, Karl. *Philipp Melanchthon als Praeceptor Germaniae.* Berlin: Hofman, 1889.

Hartmann, J., ed. *Leben und ausgewählte Schriften der Väter und Begründer der lutherischen Kirche.* (Elberfeld); part V: W. Möller on *Andreas Osiander*; part VIII: T. Pressel, *David Chytraeus.* 1862.

Hatch, Robert A. *The Boulliau Collection: An Inventory.* Philadelphia: American Philosophical Society, 1982.

Hatfield, Gary C. "Force (God) in Descartes' Physics." *Studies in the History and Philosophy of Science* 10 (1979): 113–40.

Heerbrand, Jacob. *Compendium Theologiae.* 1579.

Heilbron, John L. "Honoré Fabri, S.J., and the Accademia del Cimento." *Actes du XIIe* Congrès international d'histoire des sciences, 12 vols. Paris: Albert Blanchard, 1971: IIIB, 45–49.

Heller, Michael. "Galileo's Relativity." In *The Galileo Affair: Proceedings of the Cracow Conference*, ed. G. V. Coyne, M. Heller, and J. Zycinski. Vatican City: Specola Vaticana, 1984.

Henderson, P. A. *The Life and Times of John Wilkins.* 1910.

Hermann, Kenn. "Thomas Digges and the Elizabethan Protestant Reception of Copernicanism." Manuscript.

Hine, William L. "Mersenne and Copernicanism." *Isis* 64 (1973): 18–32.

Hobart, Michael E. *Science and Religion in the Thought of Nicolas Malebranche.* Chapel Hill: University of North Carolina Press, 1982.

Holton, Gerald. *The Thematic Origins of Scientific Thought: Kepler to Einstein*, rev. ed. Cambridge: Harvard University Press, 1988.

Hooykaas, Reijer. "Thomas Digges' Puritanism." *Archives Internationales d'Histoire des Sciences* 8 (1955): 145–49.

————. "Science and Reformation." *Journal of World History* 3 (1956): 109–39.

————. "Calvin and Copernicus." *Organon* 10:139–48.

————. "The Reception of Copernicanism in England and the Netherlands." In *The Anglo-Dutch Contribution to the Civilization of Early Modern Society*, 33–44. London, 1976.

————. "Rheticus's Treatise on Holy Scripture and the Motion of the Earth." *Journal of the History of Astronomy* 15 (1984): 77–80.

————. "The Aristotelian Background to Copernicus's Cosmology." *Journal of the History of Astronomy* 18 (1987): 111–16.

Howell, Kenneth J. "Galileo and the History of Hermeneutics." In *Facets of Faith and Science*, ed. Jitse van der Meer. University Press of America, 1995.

———. "Copernicanism and the Bible in Early Modern Science." In *Facets of Faith and Science*, ed. Jitse van der Meer. University Press of America, 1994.

Hübner, Jurgen. *Die Theologie Johannes Keplers zwischen Orthodoxie und Naturwissenschaft.* Tübingen: J. C. B. Mohr, 1975.

———. "Johannes Kepler als Geograph im Kontext des theologischen Denkens seiner Zeit." In *Wissenschaftsgeschichte um Wilhelm Schickard*, ed. Friedrich Seck. Tübingen: Mohr, 1981.

———, ed. "Der Dialog zwischen Theologie und Naturwissenschaft: Ein bibliographischer Bericht." *Forschungen und Berichte der Evangelischen Studiengenemeinschaft.* Munich: Kaiser, 1987.

Huffman, William H. *Robert Fludd and the End of the Renaissance.* London: Routledge, 1988.

Hunke, Sigrid. *Glauben und Wissen: Die Einheit europäischen Religion und Naturwissenschaft.* Düsseldorf: Econ-Verlag, 1979.

Hunter, Michael. *The Royal Society and Its Fellows, 1660–1700.* Chalfont St. Giles, England: British Society for the History of Science, 1982.

Hutchison, Keith. "Supernaturalism and the Mechanical Philosophy." *History of Science* 21 (1983): 297–333.

Iorio, Dominick. *The Aristotelians in Renaissance Italy: A Philosophical Exposition.* Lewiston, New York: Edwin Mellen Press, 1991.

Irwin, Joice. "Embryology and the Incarnation: A 16th-century debate." *Sixteenth Century Journal* 9 (3) (1978): 93–104.

Jansen, H. P. H. *Geschiedenis van de Lage Landen in jaartallen.* Utrecht: Prisma, 1988, 7th ed.

Jaki, Stanley L. *Uneasy Genius: The Life and Work of Pierre Duhem.* The Hague: Martinus Nijhoff, 1984.

Jardine, Nicholas. "Galileo's Road to Truth and the Demonstrative Regress." *Studies in the History and Philosophy of Science* 7 (1976): 277–318.

———. "The Forging of Modern Realism: Clavius and Kepler against the Sceptics." *Studies in the History and Philosophy of Science* 10 (1979): 141–74.

———. "The Significance of the Copernican Orbs." *Journal of the History of Astronomy* 13 (1982): 168–94.

———. *The Birth of History and Philosophy of Science: Kepler's A defence of Tycho against Ursus, with Essays on its Provenance and Significance.* Cambridge: Cambridge University Press, 1984.

———. "Skepticism in Renaissance Astronomy: A Preliminary Study." In *Skepticism from the Renaissance to the Enlightenment*, ed. Richard Popkin and Charles Schmitt. Wiesbaden: Otto Harrasowitz, 1987.

Jarrell, Richard A. "The Life and Scientific Work of the Tübingen Astronomer Michael Maestlin." Ph.D. thesis, University of Toronto, 1971.

———. "Maestlin's Place in Astronomy." *Physis* 17, Fasc. 1–2 (1975): 5–20.

———. "Astronomy at the University of Tübingen: The Work of Michael Maestlin." In *Wissenschaftsgeschichte um Wilhelm Schickard*, ed. Friedrich Seck. Tübingen: Mohr, 1981.

Jedin, Hubert. *The History of the Council of Trent*, 2 vols. New York: Thomas Nelson, 1957.

Jervis, Jane L. "Vogelin on the Comet of 1532: Error Analysis in the 16th Century." *Centaurus* 23 (1980): 216–29.

Johnson, Francis. "The Influence of Thomas Digges on the Progress of Astronomy in Sixteenth-Century England." *Osiris* 1 (1936): 390–410.

———. *Astronomical Thought in Renaissance England: A Study in Scientific Writings from 1500 to 1645*. Baltimore: The Johns Hopkins University Press, 1937.

Johnson, Francis, and Sanford V. Larkey. "Thomas Digges, the Copernican System, and the Idea of the Infinity of the Universe in 1576." *The Huntington Library Bulletin* 5 (1934): 69–117.

Kaiser, Christopher. "Calvin, Copernicus and Castellio." *Calvin Theological Journal* 21 (1986): 5–31.

———. "Calvin's Understanding of Aristotelian Natural Philosophy: Its Extent and Possible Origins." In *Calviniana: Ideas and Influences of Jean Calvin*, vol. 10, Sixteenth Century Essays & Studies, ed. Robert V. Schnucker, 1988.

———. *Creation and the History of Science*. Grand Rapids: Eerdmans, 1991.

Keen, Ralph. *A Melanchthon Reader*. New York: Peter Lang, 1988.

Kempfi, Andrzej. "Tolosani versus Copernicus: On a Certain Appendix to Treatise on the Truth of Holy Scripture from the Forties of the 16th Century." *Organon* 16–17 (1980–81; pub. 1983): 239–54.

Kepler, Johannes. *Johannes Kepler Gesammelte Werke*, ed. Walter von Dyck and Max Caspar. Munich: C. H. Beck, 1938–, 20 vols.

Kleiner, Scott A. "Feyerabend, Galileo, and Darwin: How to Make the Best Out of What You Have—or Think You Can Get." *Studies in the History and Philosophy of Science* 10 (1979): 285–309.

Kirch, Conrad, S.J. *Enchiridion Fontium Ecclesiasticae Antiquae*. Barcelona: Editorial Herder, 1947.

Kirsch, Irving. "Demonology and Science during the Scientific Revolution." *Journal of the History of the Behavioral Sciences* 16 (1980): 359–68.

Koenigsberger, Dorothy. *Renaissance Man and Creative Thinking: A History of Concepts of Harmony 1400–1700*. Hassocks, England: Harvester Press, 1979.

Kolb, Robert A. *Caspar Peucer's Library: Portrait of a Wittenberg Professor of the Mid-Sixteenth Century*. St. Louis: Center for Reformation Research, 1976.

————. "Historical Background of the Formula of Concord." In *A Contemporary Look at the Formula of Concord*, ed.Wilbert Rosin and Robert Preus, 12–87. St. Louis: Concordia, 1978.

————. *Confessing the Faith: Reformers Define the Church, 1530–1580*. St. Louis: Concordia, 1991.

Kowal, Charles, and Stillman Drake. "Galileo's Observations of Neptune." *Nature* 287 (1980): 311–13.

Koyré, Alexandre. *The Astronomical Revolution: Copernicus-Kepler-Borelli*, trans. R. E. W. Maddison. Paris: Hermann, 1973; New York: Dover reprint, 1992.

Knight, David. "Religion and the New Philosophy." *Renaissance and Modern Studies* 26 (1982): 147–66.

Kozhamthadam, Job. *The Discovery of Kepler's Laws: The Interaction of Science, Philosophy, and Religion*. Notre Dame: University of Notre Dame Press, 1994.

Krabbe, Otto. *Die Universität Rostock im 15. und 16. Jahrhundert*. Rostock: Adler's Erben Verlag, 1854; reprint 1970.

————. *David Chytraeus*. Rostock, 1870.

Krafft, Fritz. "Astronomie als Gottesdienst: Die Erneuerung der Astronomie Durch Johannes Kepler." In *Der Weg der Naturwissenschaft von Johannes Kepler*, ed. Gunther Hamann and Helmuth Grossing. Vienna: Verlag der Osterreichischen Akedemie der Wissenschaften, 1988.

Kraus, Hans Joachim. "Calvin's Exegetical Principles." *Interpretation* 31 (1977): 8–18.

Kren, Claudia. "Astronomical Teaching at the Late Medieval University of Vienna." *History of Universities* 3 (1983): 15–30.

Kretzmann, Norman, ed. *Infinity and Continuity in Ancient and Medieval Thought*. Ithaca: Cornell University Press, 1982.

Kretzmann, Norman, Anthony Kenny, and Jan Pinborg, eds. *The Cambridge History of Later Medieval Philosophy, from the Rediscovery of Aristotle to the Disintegration of Scholasticism, 1110–1600*. Cambridge: Cambridge University Press, 1982.

Kubrin, David C. "Providence and the Mechanical Philosophy: The Creation and Dissolution of the World in Newtonian Thought. A Study of the Relations of Science and Religion in Seventeenth Century England." Ph.D. diss., Cornell University, 1968.

Kuhn, Thomas. *The Copernican Revolution*. New York: Random House, 1959.

Kusukawa, Sachiko. *The Transformation of Natural Philosophy*. Cambridge: Cambridge University Press, 1995.

Langford, James J. *Galileo, Science and the Church*. Ann Arbor: University of Michigan Press, 1971.

Lansbergen, Jacob. *Apologia pro Commentationibus Philippi Lansbergii*. Middelburgi-Zeelandiae, 1633.

Lansbergen, Philipp. *Progymnasmatum astronomiae restitutae de motu solis*. 1619.

―――. *Bedenckingen Op den Dagelijckschen, ende Iaerlijkschen loop van den Aerdt-kloot. Mitsgaders Op de ware Af-beeldinge des sienelijcken Hemels; daer in den wonderbare Wercken Godts worden ontdeckt*. Middelburg, 1629.

Lattis, James M. "Scientific Aspects of Two Theological Questions in Thomas Aquinas." *Indiana Social Studies Quarterly* 37 (2) (1984): 24–31.

―――. *Between Copernicus and Galileo: Christopher Clavius and the Collapse of Ptolemaic Astronomy*. Chicago: University of Chicago Press, 1994.

Lerner, Lawrence S., and Edward A. Gosselin. "Galileo and the Specter of Bruno." *Scientific American* 255 (5) (1986): 126–33.

Leuven, Stedelijk Museum. *550 Jaar Universiteit Leuven*. Lembeke: J. Roegiers, 1976.

Lewis, Christopher. *The Merton Tradition and Kinematics in Late 16th and Early 17th Century Italy*. Padua: Antenore, 1980.

Lindberg, David C. "Science as Handmaiden: Roger Bacon and the Patristic Tradition." *Isis* 78 (1987): 518–36.

Lindberg, David C., and Ronald L. Numbers. "Beyond War and Peace: A Reappraisal of the Encounter between Christianity and Science." In *The Best in Theology*, vol. 1., ed. J. I. Packer, 133–49. Carol Stream, Ill.: Christianity Today, 1988.

―――, eds. *God and Nature: Historical Essays on the Encounter between Christianity and Science*. Berkeley: University of California Press, 1986.

Lindberg, David C., and Robert S. Westman. *Reappraisals of the Scientific Revolution*. Cambridge: Cambridge University Press, 1990.

de Lubac, Henri. *Exégèse Mediévale: Les Quatre Sens de l'Ecriture*. Paris: Aubier, 1959.

Lueker, Erwin L., ed. *Lutheran Cyclopedia*. St. Louis: Concordia, 1975. Rev. ed.

Luther, Martin. "On the Councils and the Church" (1539). In *Luther's Works: Church and Ministry III*, vol. 41, ed. Eric W. Gritsch. Philadelphia: Fortress Press, 1966.

―――. "Against the Papacy, an Institution of the Devil." In *Luther's Works: Church and Ministry III*, vol. 41, ed. Eric W. Gritsch. Philadelphia: Fortress Press, 1966.

Maestlin, Michael. *Epitome Astronomiae: qua brevi explicatione omnia, tam ad sphaericam quam theoricam eius partem pertinentia, ex ipsius scientiae fontibus deducta.* Tübingen: Gruppenbacchius, 1597.

Makin, W. "The Philosophy of Pierre Gassendi: Science and Belief in 17th-Century Paris and Provence." Ph.D. thesis, Open University, UK, 1986.

Malherbe, Jean-François. *La langage théologique à l'âge de la science: Lecture de Jean Ladrière.* Paris: Cerf, 1985.

Mandrou, Robert. *From Humanism to Science, 1480 to 1700,* trans. Brian Pearce. Hassocks, England: Harvester Press, 1979.

Mara, Maria Grazia. "Ambrose of Milan." In *Patrology,* vol. 4, ed. Angelo de Beradino. Westminster, Md.: Christian Classics, 1991.

Marcel, Pierre. "Calvin et la science: Comment on fait l'histoire." *Revue reformée* 69 (1966): 51.

———. "Calvin et Copernic: La legende ou les faits? La science et l'astronomie chez Calvin." *Revue reformée* 31 (1980): 1–210.

Marion, Jean-Luc. *Sur le prisme metaphysique de Descartes: Constitution et limites de l'onto-theo-logie dans la pensée cartésienne.* Paris: Presses Universitaires de France, 1986.

Martin, R. N. D. "Saving Duhem and Galileo: Duhemian Methodology and the Saving of the Phenomena." *History of Science* 25 (1987): 301–19.

———. *Pierre Duhem: Philosophy and History in the Work of a Believing Physicist.* La Salle, Ill.: Open Court, 1991.

Massing, Jean-Michel. "A 16th Century Illustrated Treatise on Comets." *Journal of the Warburg and Courtauld Institutes* 40 (1977): 318–22.

Mason, Stephen F. "The Scientific Revolution and the Reformation." *A History of the Sciences,* 175–91. New York: Collier Books, 1962.

Maurer, W. *Der Junge Melanchthon zwischen Humanismus und Reformation.* Göttingen: Vaudenhoeck and Ruprecht, 1969.

McClaughlin, Trevor. "Censorship and Defenders of the Cartesian Faith in Mid-17th Century France." *Journal of the History of Ideas* 40 (1979): 563–81.

McClumpha, B. "Some Aspects of the Intellectual Relations between Galileo and the Jesuits." Ph.D. thesis, Leeds University, UK, 1985.

McColley, Grant. "An Early Friend of the Copernican Theory: Gemma Frisius." *Isis* 26 (1936): 322–25.

———. "The Defense of Galileo." *Smith College Studies in History* 22 (3–4) (1937). Northhampton, Mass.

———. "The Debt of Bishop John Wilkins to the *Apologia pro Galileo* of Tommaso Campanella." *Annals of Science* 4 (1939): 150–68.

McColley, Grant, and John H. Randall. "Comments and Criticisms" (concerning Campanella and Galileo). *Journal of Philosophy* 36 (1939): 157–58.

McMullin, Ernan. "Introduction: Galileo, Man of Science." In *Galileo, Man of Science*. New York: Basic Books, 1967.

———. "Bruno and Copernicus." *Isis* 78 (1987): 55–74.

———. "Conceptions of Science in the Scientific Revolution." In *Reappraisals of the Scientific Revolution*, ed. David C. Lindberg and Robert S. Westman, 27–92. Cambridge: Cambridge University Press, 1990.

McNorris, Michael N. "Science as Scientia." *Physis* 23 (1981): 171–96.

Meattini, Valerio. "Copernico: Un 'caso' per la storia della scienza." *Critica Storica* 16 (1979): 623–45.

Meinel, Christoph. "Early 17th-century Atomism: Theory, Epistemology, and the Insufficiency of Experiment." *Isis* 79 (1988): 68–103.

Mela, Pomponius. *Cosmographia sive de situ orbis* (late 15th century).

Melancthon, Philip. "Initia Doctrinae Physicae." In *Opera Omnia*, ed. Carl Bretschneider, 656–57. *Corpus Reformatorum*, vol. 13 (1834).

———. *Loci Communes*, trans. J. A. O. Preus. St. Louis: Concordia, 1992.

———. *A Melanchthon Reader*, trans. Ralph Keen. New York: Peter Lang, 1988.

———. *Commentary on Romans*, trans. Fred Kramer. St. Louis: Concordia, 1992.

Merton, Robert K. "Science, Technology, and Society in Seventeenth Century England." *Osiris* 38 (1937): 360–632.

Migne, J. P. *Patrologia Latina*. Paris: Garnier, 1878.

Moesgaard, Kristian P. "Copernican Influence on Tycho Brahe." In *Colloquia Copernicana I, in Studia Copernicana V*. Warsaw: Polish Academy of Sciences, 1972.

Moore, James R. *The Post-Darwinian Controversies: A Study of the Protestant Struggle to Come to Terms with Darwin in Great Britain and America, 1870–1900*. New York: Cambridge University Press, 1979.

Moran, Bruce T. "The Universe of Philip Melancthon: Criticism and Use of the Copernican Theory." *Comitatus* 4 (1973): 1–23.

Moran, Bruce T., and Christoph Rothmann. "The Copernican Theory and Institutional and Technical Influences on the Criticism of Aristotelian Cosmology." *Sixteenth Century Journal* 13 (3) (1982): 85–103.

Moraze, Charles. *Les origines sacrées des sciences modernes*. Paris: Fayard, 1986.

Moss, Jean Deitz. "Galileo's *Letter to Christina*: Some Rhetorical Considerations." *The Renaissance Quarterly* 36 (1983): 547–76.

———. "Galileo's Rhetorical Strategies in Defense of Copernicanism." In *Novita Celesti e Crisi del Sapere*, ed. P. Galluzzi, 95–103. Florence: Giunti Barbera, 1984.

———. "The Rhetoric of Proof in Galileo's Writings on the Copernican System." In *Reinterpreting Galileo Studies in Philosophy and the History of*

Philosophy, ed. William A. Wallace, 179–204. Washington: Catholic University Press, 1986.

———. *Novelties in the Heavens*. Chicago: University of Chicago Press, 1993.

Müller, Gerhard. "Philipp Melanchthon zwischen Pädagogik und Theologie." In *Humanismus im Bildungswesen des 15. und 16. Jahrhunderts*, ed. Wolfgang Reinhard. Weinheim: Acta Humaniora, 1984.

Müller-Jahncke, and Wolf-Dieter. *Astrologisch-magische: Theorie und Praxis in der Heilkunde der fruhen Neuzeit*. Stuttgart: Steiner, 1985.

Multhauf, Robert P. "Copernicus and Bacon as Renovators of Science." In *Science and History: Studies in Honor of Edward Rosen*, 489–99. Studia Copernicana 16. Wroclaw: Ossolineum, 1978.

Naylor, Ronald H. "Galileo's Theory of Projectile Motion." *Isis* 71 (1980): 550–70.

———. "Mathematics and Experiment in Galileo's New Sciences." *Annali dell' Instituto e Museo di Storia della Scienza di Firenze* 4 (1): 55–63.

———. "The Role of Experiment in Galileo's Early Work on the Law of Fall." *Annals of Science* 37 (1980): 363–78.

———. "Galileo's Early Experiments on Projectile Trajectories." *Annals of Science* 40 (1983): 391–94.

Neher, Andre. *Jewish Thought and the Scientific Revolution of the 16th Century*. Oxford: Oxford University Press for the Littman Library, 1986.

Neugebauer, Otto. "The Equivalence of Ptolemaic and Copernican Astronomy." In *Vistas in Astronomy*. 1968.

Newman, William R. "Boyle's Debt to Corpuscular Alchemy." In *Robert Boyle Reconsidered*, ed. Michael Hunter. Cambridge: Cambridge University Press, 1994.

Nier, Keith A. "The Importance of Historical Accuracy in Philosophy of Science: The Case of Curd's Conception of Copernican Rationality." *Philosophy of Science* 53 (1986): 372–94.

Nobis, Heribert M. "Die Vorbereitung der copernicanischen Wende in der Wissenschaft der Spätscholastik." In *Mathemata: Festschrift für Helmuth Gericke*, ed. M. Folkerts and U. Lindgren. Stuttgart: Steiner, 1985.

Novak, David, and Norbert Samuelson, eds. *Creation and the End of Days: Judaism and Scientific Cosmology*. Proceedings of the 1984 Meeting of the Academy of Jewish Philosophy. Lanham: University Press of America, 1986.

Oberman, Heiko A. "Reformation and Revolution: Copernicus' Discovery in an Era of Change." In *The Dawn of the Reformation: Essays in Late Medieval and Early Reformation Thought*, 179–203. Edinburgh: Clark, 1986.

Olivieri, Luigi, ed. *Aristotelismo Veneto e Scienza moderna: Atti del 250 anno accademico del Centro per la Storia della Tradizione Aristotelica nel Veneto*. Padua: Antenore, 1983.

Origen of Alexandria. *Contra Celsum*, ed. Henry Chadwick. Cambridge: Cambridge University Press, 1965.

Osler, Margaret J. "Descartes and Charleton on Nature and God." *Journal of the History of Ideas* 40 (1979): 445–56.

———. "Eternal Truths and the Laws of Nature: The Theological Foundations of Descartes' Philosophy of Nature." *Journal of the History of Ideas* 46 (1985): 349–62.

———. *Divine Will and the Mechanical Philosophy*. Cambridge: Cambridge University Press, 1994.

Osler, Margaret J., and Paul Lawrence Farber, eds. *Religion, Science, and Worldview: Essays in Honor of Richard S. Westfall*. Cambridge: Cambridge University Press, 1985.

Patterson, Louise Diehl. "Leonard and Thomas Digges: Biographical Notes." *Isis* 42 (1951): 120–21.

Pedersen, Olaf. "Galileo and the Council of Trent: The Galileo Affair Revisited." *Journal of the History of Astronomy* 14 (1983): 1–29.

———. *Tycho Brahe og astronomiens genfodsel*. Arhus: Foreningen Vidensk-abskhistorisk Museums Venner, 1985.

Pelseneer, Jean. "L'origine protestante de la science moderne." *Lychnos* (1946–47): 246–48.

Pesce, Mauro. "Il *Consensus Veritatis* di Christoph Wittich e la distinzione tra verità scientifica e verità biblica." *Annali di storia dell'esegesi* 9 (1992): 53–76.

Pettegree, Andrew. *Emden and the Dutch Revolt: Exile and the Development of Reformed Protestantism*. Oxford: Clarendon Press, 1992.

Peucer, Caspar. *Elementa doctrinae de circulis coelestibus, et primo motu, recognita et correcta*. Wittenberg, 1553.

———. *Commentarius de praecipuis divinationum generibus*. Wittenberg, 1553.

Pizzorno, Benedetto. "Indagine sulla logica Galileiana." *Physis* 21 (1979): 71–102.

Polman, A. D. R. *The Word of God According to St. Augustine*. Grand Rapids: Eerdmans, 1961.

Popkin, Richard H. "The Religious Background of 17th-Century Philosophy." *Journal of the History of Philosophy* 25 (1987): 35–50.

Poupard, Paul, ed. *Galileo Galilei 360 anni di storia, 1633–1983*. Rome: Piemme, 1984.

———. *Galileo Galilei: Toward a Resolution of 350 Years of Debate, 1633–1983*. Pittsburgh: Duquesne University Press, 1987.

Preus, Robert D. *The Theology of Post-Reformation Lutheranism*. 2 vols. St. Louis: Concordia, 1970.

Price, J. L. *Holland and the Dutch Republic in the Seventeenth Century: The Politics of Particularism*. Oxford: Clarendon Press, 1994.

Principe, Lawrence. "Boyle's Alchemical Pursuits." In *Robert Boyle Reconsidered*, ed. Michael Hunter. Cambridge: Cambridge University Press, 1994.

Ptolemy, Claudius. *Cosmographia Tabulae*. Leicester, England: Magna Books, 1990.

Raeder, Hans, Elis Strömgren, and Bengt Strömgren. *Tycho Brahe's Description of His Instruments and Scientific Work as Given in Astronomiae Instauratae Mechanica*. Copenhagen: Det Kongelige danske Videnskabernes Selksab, 1946.

Ratner, Joseph. "Some Comments on Rosen's 'Calvin's Attitude toward Copernicus.'" *Journal of the History of Ideas* 22 (1961).

Redondi, Pietro. *Galileo Heretic*, trans. Raymond Rosenthal. Princeton: Princeton University Press, 1987.

Reid, W. Stanford. "Calvin and the Founding of the Geneva Academy." *Westminster Theological Journal* 18 (1955): 1–33.

Reitsma, J. *Geschiedenis van der Hervorming en de Hervormde Kerk*. Utrecht, 1933.

Rheticus, Georg Joachim. *Rheticus's Treatise on Holy Scripture and the Motion of the Earth*, ed. and trans. Reijer Hooykaas. Amsterdam: North-Holland Publishing, 1984.

———. "The First Narrative." In *Three Copernican Treatises*, ed. and trans. Edward Rosen. New York: Dover Publications, 1939.

Richgels, Robert W. "Scholasticism Meets Humanism in the Counter-Reformation: The Clash of Cultures in Robert Bellarmine's Use of Calvin in the *Controversies*." *Sixteenth Century Journal* 6 (1975): 53–66.

Robbins, Frank E. "The Hexaemeral Literature." Ph.D. thesis, University of Chicago, 1912.

Roberts, Lawrence D., ed. *Approaches to Nature in the Middle Ages*. Binghamton, New York: Center for Medieval and Early Renaissance Studies, 1982.

Romagnoli, Maria. "Due Lettere inedite di Galileo Galilei." *Annali dell'Istituto e Museo di Storia di Scienza di Firenze* 7 (1): 29–34.

Rosen, Edward. "Calvin's Attitude toward Copernicus." *Journal of the History of Ideas* 21 (1960): 431–41.

———. "Rheticus, Georg Joachim." In the *Dictionary of Scientific Biography*, vol. 11, 395–98. Washington D.C.: American Council of Learned Societies, 1975.

———. "Kepler and the Lutheran Attitude Towards Copernicanism in the Context of the Struggle Between Science and Religion." *Vistas in Astronomy* 18 (1975): 317–37.

————. "Render Not Unto Tycho That Which Is Not Brahe's." *Sky and Telescope* 61 (1981): 476–77.

————. *Copernicus and the Scientific Revolution*. Malabar, Fla.: Krieger, 1984.

————. "The Dissolution of the Solid Celestial Spheres." *Journal of the History of Ideas* 46 (1985): 13–31.

————. *Three Imperial Mathematicians*. New York: Abaris Books, 1986.

Rossi, Paolo. "Galileo Galilei e li libro dei Salmi." *Rivista della Filosophia* 69 (1978): 45–71.

Rougier, Louis. *Astronomie et Religion en Occident*. Paris: Presses Universitaires de France, 1980.

Rupp, E. Gordon. "Philipp Melanchthon." In *A History of Christian Doctrine*, ed. Hubert Cutliffe-Jones, 373–78. Philadelphia: Fortress Press, 1978.

Russell, John. "The Copernican System in Great Britain." In *The Reception of Copernicus' Heliocentric Theory*, ed. Jerzy Dobrzycki. Dordrecht: D. Reidel, 1972.

Russo, François. "Pour une histoire de la conception des types généraux de systèmes et de processus de la nature." *Archives de Philosophie* 46 (1983): 105–28.

Rybka, Eugeniusz. "Copernican Ideas in Kepler's 'Epitome Astronomiae Copernicanae.'" In *Science and History: Studies in Honor of Edward Rosen*, 461–72. Studia Copernicana, 16. Wroclaw: Ossolineum, 1978.

Santillana, Giorgio de. *The Crime of Galileo*. Chicago: University of Chicago Press, 1955.

Schaffer, Simon. "Godly Men and Mechanical Philosophers: Souls and Spirits in Restoration Natural Philosophy." *Science in Context* 1 (1987): 55–85.

Schales, Jurgen. "Zur Frage der Erkenntnis bei Galileo Galilei." In *Nebenwege der Naturphilosophie und Wissenschaftsgeschichte*, ed. Gottfried Heinemann. (Kassel: Gesamthochschule Kassel, 1987.

Scheurleer, Th. H. Lunsingh, and G. H. M. Posthumus Meyjes. *Leiden University in the Seventeenth Century*. Leiden: E.J. Brill, 1975.

Schilling, Heinz. "Die Konfessionalisierung im Reich, Religiöser und gesellschaftlicher Wandel in Deutschland zwischen 1555 und 1620." *Historische Zeitschrift* 246 (1988): 1–45.

————. "Reformation und Konfessionalisierung in Deutschland und die neuere deutsche Geschichte." *Gegenwartskunde, Gesellschaft, Staat, Erziehung* Sonderheft 6 (1988): 11–29.

Schmitt, Charles B. *Aristotle and the Renaissance*. Cambridge: Harvard University Press, 1983.

Schofield, Christine. *Tychonic and Semi-Tychonic World Systems*. New York: Springer, 1981.

Segonds, Alain. "Tycho Brahe et L'Alchimie." In *Alchimie et Philosophie à la Renaissance*, ed. Jean Claude Margolin and Sylvain Matton. Paris: Libraire Philosophique, J. Vrin, 1993.

Shakelford, Jole. *Paracelsianism in Denmark and Norway in the Sixteenth and Seventeenth Centuries*. Ann Arbor: University Microfilms International, 1989.

———. "Paracelsianism and Patronage in Early Modern Denmark." In *Patronage and Institutions: Science, Technology, and Medicine at the European Court, 1500–1750*, ed. Bruce T. Moran. The Boydell Press, 1991.

———. "Tycho Brahe, Laboratory Design, and the Aim of Science." *Isis* 84 (1992): 211–30.

Shapere, Dudley. *Galileo: A Philosophical Study*. Chicago: University of Chicago Press, 1974.

Shapin, Steven. *A Social History of Truth, Civility, and Science in Seventeenth-Century England*. Chicago: University of Chicago Press, 1994.

Shapin, Steven, and Simon Schaffer. *The Leviathan and the Air-Pump: Hobbes, Boyle and the Experimental Life*. Princeton: Princeton University Press, 1985.

Shapiro, Barbara. *John Wilkins, 1614–1672: An Intellectual Biography*. Berkeley: University of California Press, 1969.

———. *Probability and Certainty in 17th Century England: A Study of the Relationships between Natural Science, Religion, History, Law, and Literature*. Princeton: Princeton University Press, 1983.

———. "Early Modern Intellectual Life: Humanism, Religion, and Science in Seventeenth Century England." *History of Science* 29 (1993): 45–71.

Shea, William R. "Marin Mersenne: Galileo's 'traduttore-traditore.'" *Annali dell'Istituto e Museo di Storia di Scienza di Firenze* 2 (1) (1977): 55–70.

———. "Galileo and the Justification of Experiments." In *Historical and Philosophical Dimensions of Logic, Methodology, and Philosophy of Science*, ed. Robert E. Butts and Jaakko Hintikka, 81–92. Dordrecht: Reidel, 1977.

———. "Descartes and the Rosicrucians." *Annali dell'Istituto e Museo di Storia di Scienza di Firenze* 4 (2) (1979): 29–47.

———. "Melchior Inchofer's *Tractatus Syllepticus*: A Consultor of the Holy Office Answers Galileo." In *Novita Celesti e Crisi del Sapere*, ed. Paolo Galluzzi, 284–92. Florence: Giunti Barbera, 1984.

Smith, A. Mark. "Galileo's Proof for the Earth's Motion from the Movement of Sunspots." *Isis* 76 (1985): 543–51.

Snelders, H. A. M. "Science and Religion in the Seventeenth Century: The Case of Northern Netherlands." In *Italian Scientists in the Low Countries in the XVIIth and XVIIIth Centuries*, ed. C. S. Maffioli and L. C. Palm, 65–77. Amsterdam: Rodopi. 1989.

Snelders, H. A. M., and K. van Berkel, eds. *Natuurwetenschappen van Renaissance tot Darwin.* Den Haag: Martinus Nijhoff, 1981.

Southgate, B. C. "A Philosophical Divinity: Thomas White and Aspects of Mid-17th Century Science and Religion." *History of European Ideas* 8 (1987): 45–59.

Sprat, Thomas. *The History of the Royal Society.* London, 1667.

Stadler, Michael. "Unendliche Schöpfung als Genesis von Bewusstsein: überlegungen zur Geistphilosophie Giordano Brunos." *Philosophisches Jahrbuch* 93 (1986): 39–60.

Stauffer, Richard. "Calvin et Copernic." *Revue de l'histoire des religions* 179 (1971): 31–40.

Stengers, Jean. "Les premiers Coperniciens et la Bible." *Revue Belge de la Philologie et Histoire* 62 (1984): 703–19.

Stephenson, Bruce. *Kepler's Physical Astronomy.* New York: Springer-Verlag, 1987.

Stevens, H. N. *Ptolemy's Geography. A Brief Account of the Printed Editions Down to 1730.* London, 1908.

Stevin, Simon. *The Principal Works of Simon Stevin.* Amsterdam: C.V. Swets & Zeitlinger, 1955, vols. 1 and 3.

Stimson, Dorothy. *The Gradual Acceptance of the Copernican Theory of the Universe.* New York: Baker & Taylor, 1917.

———. "Puritanism and the New Philosophy in Seventeenth-Century England." *Bulletin of the Institute of the History of Medicine* 3 (1935): 321–34. Reprinted in *Puritanism and the Rise of Modern Science,* ed. I. Bernard Cohen, 151–58. New Brunswick: Rutgers University Press, 1990.

Strubberg, Johann. *Index theologorum Evangelicorum-lutheranorum chronologicus.* Manuscript in library of Indiana University.

Struik, Dirk. *The Land of Stevin and Huygens: A Sketch of Science and Technology in the Dutch Republic during the Golden Century.* Dordrecht: D. Reidel, 1981.

Stupperich, Robert. *Melanchthons Werke in Auswahl,* 5 vols. Gütersloh: C. Bertelsmann, 1951.

Swerdlow, Noel M. "Pseudodoxia Copernicana." *Archives internationales d'histoire des sciences* 26 (1976): 108–58.

Tanner, Norman. *Decrees of the Ecumenical Councils.* Washington, D.C.: Georgetown University Press, 1992.

Tarabochia, Alessandra. *Esegesi biblica e cosmologia note sull'interpretazione patristica e mediovale di Genesi 1, 2.* Milan: Universita Cattolica del Sacro Cuore, 1981.

Taub, Liba. *Ptolemy's Universe: The Natural Philosophical and Ethical Foundations of Ptolemy's Astronomy.* Chicago: Open Court, 1993.

Thijssen-Schoute, C. Louise. "Le Cartesianisme aux Pays-Bas." In *Descartes et le Cartésianisme hollandais: études and documents,* ed. E. J. Diksterhuis et al. Presses universitaires de France, 1950.

———. *Nederlands Cartesianisme.* Amsterdam: Noord-Holland Uitgevers, 1954.

Thoren, Victor E. "Tycho Brahe as the Dean of a Renaissance Research Institute." In *Religion, Science, and Worldview: Essays in Honor of Richard S. Westfall,* ed. M. J. Osler and P. L. Farber. Cambridge: Cambridge University Press, 1985.

———. *The Lord of Uraniborg: A Biography of Tycho Brahe.* Cambridge: Cambridge University Press, 1991.

Thuillier, Pierre. "Les Jesuites ont-ils été des pionniers de la science moderne." *Recherche* 19 (1988): 88–92.

Torrance, Thomas F. *The Hermeneutics of John Calvin.* Monograph supplements to the *Scottish Journal of Theology.* Scottish Academic Press, 1988.

Torrini, Maurizio. "Et vidi coelum novum et terram novam: A proposito di rivoluzione scientifica e libertinismo." *Nuncius* 1 (2) (1986): 49–77.

van der Wall, Ernestine. "Orthodoxy and Skepticism in the Early Dutch Enlightenment." In *Skepticism and Irreligion in the Seventeenth and Eighteenth Centuries,* ed. Richard Popkin and Arjo Vanderjagt. Leiden: E. J. Brill, 1993.

Van Helden, Albert. "Galileo on the Sizes and Distances of the Planets." *Annali dell'Istituto e Museo di Storia della Scienza di Firenze* 7 (2) (1982): 65–86.

Van Winden, J. C. M. *"Idee" en "Materie" in de vroeg-christlijke uitleg van de beginwoorden van genesis.* Amsterdam: Noord-Hollandssche Uitgevers Maatschappij, 1985.

Vasoli, Cesare. "Tradizione e Nuovo Scienza." In *Novita Celesti e Crisi del Sapere,* ed. Paolo Galluzzi, 73–94. Florence: Giunti Barbera, 1984.

Velthuysen, Lambertus van. *Opera Omnia.* Rotterdam, R. Leers, 1680.

Verbeek, Theo. *Descartes and the Dutch Early Reactions to Cartesian Philosophy, 1637–1650.* Carbondale: Southern Illinois University Press, 1992.

Verbeke, Gerard. *The Presence of Stoicism in Medieval Thought.* Catholic University of America Press, 1983.

Verhoef, Pieter A. "Luther's and Calvin's Exegetical Library." *Calvin Theological Journal* 13 (1968): 5–20.

Vermij, Rienk. *Secularisering en Natuurwetenschap in de seventiende en achtentiende eeuw: Bernard Nieuwentijt.* Amsterdam: Rodopi, 1991.

———. "Het copernicanisme in de Republiek: een verkenning." *Tijdschrift der geschiednis.* 1993.

Vigano, Mario. *Il Mancato Dialogo tra Galileo e i teologi.* Rome: La Civilta Cattolica, 1969.

Vinaty, Bernard. *Galileo Galilei: 350 ans d'histoire, 1633–1983.* Tournai: Desclée International, 1983.

———. "Galilei: Lo scandalo della ragione: The scandal of reason." *Scientia* 117 (1982): 147–353.

Voetius, Gisbertus. *Thersites heautontimorumenos* ex officina Abrahami ab Herwyck et Hermanni Ribbii, 1635.

———. *Godzaligheid te verbinden met de wetenschap: inaugurele reden gehouden aan de Illustre School te Utrecht op 21ste augustus 1634.*

Vrijer, M. J. A. de. *Henricus Regius, en "Cartesiaansch" hoogleraar aan de Utrechtsche Hoogeschool.* The Hague, 1917.

Wallace, William A. *Galileo and His Sources: The Heritage of the Collegio Romano in Galileo's Science.* Princeton: Princeton University Press, 1984.

———. "Galileo's Early Arguments for Geocentrism and His Later Rejection of Them." In *Novita Celesti e Crisi del Sapere,* ed. Paolo Galluzzi, 31–40. Florence: Giunti Barbera, 1984.

———. "The Certitude of Science in Late Medieval and Renaissance Thought." *History of Philosophy Quarterly* 3 (1986): 281–91.

———, ed. *Reinterpreting Galileo Studies in Philosophy and the History of Philosophy,* 15. Washington, D.C.: Catholic University of America Press, 1986.

Wardeska, Zofia. "Copernicus und die deutschen Theologen des 16. Jahrhunderts." In *Nikolaus Copernicus zum 500. Geburtstag,* ed. F. Kaulbach, U. W. Bargenda and J. Bluhdorn. Cologne: Boehlau Verlag, 1973.

Waterbolk, E. H. "The Reception of Copernicus's Teachings by Gemma Frisius (1508–1555)." *Lias* 1 (1974): 225–42.

Weisheipl, James A. *Nature and Motion in the Middle Ages,* ed. William E. Carroll. Washington, D.C.: Catholic University of America Press, 1985.

Westfall, Richard S. *The Construction of Modern Science: Mechanisms and Mechanics.* Cambridge: Cambridge University Press, 1971.

———. *Never at Rest: A Biography of Isaac Newton.* Cambridge: Cambridge University Press, 1980.

———. "Newton's Marvelous Years of Discovery and Their Aftermath: Myth versus Manuscript." *Isis* 71 (1980): 109–21.

———. "Galileo and the Accademia dei Lincei." In *Novita Celesti e Crisi del Sapere,* ed. Paolo Galluzzi, 189–200. Florence: Giunti Barbera, 1984.

———. "The Influence of Alchemy on Newton." In *Mapping the Cosmos,* ed. Jane Chance and R. O. Wells, Jr., 98–117. Houston: Rice University Press, 1985.

———. "Scientific Patronage: Galileo and the Telescope." *Isis* 76 (1985): 18–22.

———. "The Rise of Science and the Decline of Orthodox Christianity: A Study of Kepler, Descartes and Newton." In *God and Nature: Historical Essays on the Encounter between Christianity and Science,* ed. David C. Lindberg and Ronald L. Numbers, 219–24. Berkeley: University of California Press, 1986.

Westman, Robert S. "The Comet and the Cosmos: Kepler, Mästlin and the Copernican Hypothesis." *Colloquia Copernicana I,* in *Studia Copernicana V.* Warsaw: Polish Academy of Sciences, 1972.

———. "The Melanchthon Circle, Rheticus, and the Wittenberg Interpretation of the Copernican Theory." *Isis* 66 (1975): 165–93.

———. "The Wittenberg Interpretation of the Copernican Theory." In *The Nature of Scientific Discovery,* ed. Owen Gingerich. Washington, D.C.: The Smithsonian Institution, 1975.

———. "Humanism and Scientific Roles in the 16th Century." In *Humanismus und Naturwissenschaften,* ed. R. Schmitz and F. Krafft, 83–99. Boppard am Rhein: Boldt, 1980.

———. "The Astronomer's Role in the Sixteenth Century: A Preliminary Study." *History of Science* 23 (1980): 105–47.

———. "The Copernicans and the Churches." In *God and Nature: Historical Essays on the Encounter between Christianity and Science,* ed. David C. Lindberg and Ronald L. Numbers. Berkeley: University of California Press, 1986.

———, ed. *The Copernican Achievement.* Berkeley/Los Angeles: University of California Press, 1975.

White, Andrew D. *The History of the Warfare of Science with Theology in Christendom.* 2 vols. New York: Appleton, 1896.

White, Robert. "Calvin and Copernicus: The Problem Reconsidered." *Calvin Theological Journal* 15 (1980): 233–43.

Wilkins, John. *The Discovery of a New World.* London, 1638.

Williams, Arnold. *The Common Expositor.* Chapel Hill: University of North Carolina Press, 1948.

Wilson, Curtis. "Horrocks, Harmonies, and the Exactitude of Kepler's Third Law." In *Science and History: Studies in Honor of Edward Rosen,* 235–59. Studia Copernicana, 16. Wroclaw: Ossolineum, 1978.

Wisan, Winifred. "Galileo and the Process of Scientific Creation." *Isis* 75 (1984): 269–86.

———. "On the Chronology of Galileo's Writings." *Annali dell'Istituto e Museo di Storia della Scienza di Firenze* 9 (2): 85–88.

———. "Galileo and God's Creation." *Isis* 77 (1986): 473–86.

Wittich, Christoph. *Dissertationes duae, una de usu et abusu scripturae in rebus philosophicis, altera de Cartesii sententia de quiete terrae.* 1653.

———. *Consideratio theologica de stylo scripturae quem adhibet cum rebus naturalibus sermonem instituit.* 1656.

————. *Consensus veritatis in Scriptura divina et infallibili revelatae cum veritate philosophica a Renato des Cartes detecta.* Nijmegen, 1659.

Wittich, Christian. *Theologica Pacifica.* 3rd ed.

Wolf, Rudolf. *Geschichte der Astronomie.* Munich: Oldenbourg, 1877.

Wootton, David. *Paolo Sarpi: Between Renaissance and Enlightenment.* Cambridge: Cambridge University Press, 1983.

Zaiser, Hedwig. *Kepler als Philosoph.* Stuttgart: E. Suhrkamp, 1932.

Index